Francis Cruger Moore

How to Build Fireproof and Slow Burning

Francis Cruger Moore

How to Build Fireproof and Slow Burning

ISBN/EAN: 9783337249243

Printed in Europe, USA, Canada, Australia, Japan

Cover: Foto ©berggeist007 / pixelio.de

More available books at **www.hansebooks.com**

HOW TO BUILD FIREPROOF

AND

SLOW-BURNING

BY

FRANCIS C. MOORE

President of the Continental
Fire Insurance Co.; Author
of How to Build a Home;
Water Works and Pipe Distribution; Fire Insurance and
Causes of Fires, Etc., Etc.

THIRD EDITION
1899

NEW YORK
CONTINENTAL PRINT

PREFACE.

BY way of preface to the following treatise, I wish to explain that it has been prepared after careful consultation with well-known experts and after observation and study of numerous fires in "fireproof" structures, especially of those which caused losses to my own company. I offer it to property owners who contemplate building, feeling that I can, without immodesty, claim that the suggestions are important and worthy of consideration for the reason stated—that they have been revised by competent judges and have already run the gauntlet of expert criticism.

F. C. M.

New York, February 1899.

HOW TO BUILD FIREPROOF,

BY

FRANCIS C. MOORE,

President of the Continental Insurance Co., N. Y.
Delegate of New York Board of Fire Underwriters to the Board
of Examiners of the N. Y. B'ld'g Dep't.

———◆———

IT seems advisable, in a treatise of this kind, to state, as premises, certain propositions which might be treated as deductions. Some of them are axiomatic or self-evident, needing no demonstration, and ought to appeal to any practical mind as being truths, rather illustrated than demonstrated by the experience of the past few years. In conformity with this line of treatment, I desire to state by way of premise:

First.—It may be claimed that no construction is fireproof, and that even iron and masonry could with propriety be designated as "slow burning." The iron or steel used in a modern building has, in its time, been smelted in a furnace which presented no greater capacity for running metal into pigs than some of our modern buildings, whose interior openings from cellar to roof correspond to the chimney of a furnace and the front door to its tuyere. Indeed if a pyrometer could be adjusted during the progress of a fire it would be found to rise quite as high as in any forge.

Second.—GLASS WINDOWS will not prevent the entrance of flame or heat from a fire in an exposing building. It may seem strange that so obvious a proposition should be thought worth stating, and yet to-day more than 75% of the "fireproof" structures of the country have window openings to the extent of from 40% to 75% of the superficial area of each enclosing wall without fireproof shutters. Heat from a building across a wide street finds ready entrance through windows and the several fireproof floors serve only to hold ignitible merchandise in the most favorable form of distribution for ignition and combustion, like a great gridiron, to the full force of an outside fire. This was the case in the burning of the Manhattan Bank Building, on Broadway, in New York, and of the Horne Building, in Pittsburg. The latter building was full of plate glass windows, 16x16 feet. Such buildings are not more capable of protecting their contents than a glass show-case would be. A recent article on the Pittsburg fire in the "Engineering News" aptly expresses this in the following words: "There seems to be some irony in calling buildings fireproof which opposed hardly anything to a fire from across the street more sturdy than plate glass!"

Third.—OPENINGS THROUGH FLOORS for stairways or elevators, and for gas, water and steam pipes and electric wires, from floor to floor of fireproof buildings, tend to the spread of flame like so many flues and should be fire-stopped at each story. This fault is more generally overlooked than any other. Ducts for piping, wiring, etc., should never be of wood. In the Mills Building, in New York, a fire, not long since, jumped through two or three floors from the one on which it originated, by means of the passageways for piping, electric wiring, etc., comparatively small ducts, but sufficient for the passage of flame. In one instance, the fire skipped one floor, where it was cut off, and ignited the second floor above.

Fourth.—In view of the fact that it is necessary to cover iron with non-combustible, non-conducting material to prevent its exposure to fire and consequent expansion, and in view of the fact that all iron-work, except cast-iron, will rust to the point of danger, it is best to use cast-iron for all vertical supports, columns, pillars, etc. It is not advisable, of course, to have floor beams of cast-iron (except in the form of Hodgkinson beams thoroughly tested.) If a floor beam should give way, however, it might not necessarily wreck the building, whereas if a vital column should give

Fireproofing Iron Members

way a collapse of the entire structure might result.

At a convention held some years ago in New York, at which were present a greater number of experts in iron than probably ever met before or since in one room, there was not one who contended that cast-iron would rust beyond the harmless incrustation of the thickness of a knife blade, whereas there was not one who did not believe wrought iron would rust to the point of danger; and there was not one who claimed to know whether steel would or not, each admitting that steel had not been sufficiently tested as to rust to warrant a reliable opinion. If it could be relied upon as rust proof, it would be superior to all other material for fire-proof buildings because of its great strength in proportion to weight. The use of steel in construction is growing, because it is cheaper than wrought-iron, as lighter weights are used for the same strength, but while in some respects superior to wrought iron, some of the prevailing impressions with regard to it are erroneous. Defects not possible of detection by tests are liable to exist in its structure. Among the first steel beams brought to the city of New York there were instances in which they were actually broken in two by falling from the level of trucks to the pavement, probably due to their having been rolled when too cold, as steel when rolled below a certain temperature becomes brittle. **Better beams are now made.**

Marginal note: Rust

In my opinion, cast-iron colums are superior to steel and more reliable. It is not generally known that American cast-iron is vastly superior to English cast-iron, and will stand a greater strain without breaking. Cast-iron, moreover, will not expand under heat to the same extent as wrought iron and steel, which is another fact in its favor.

No bearing column should be placed in such a position that it cannot be uncovered and exposed for examination without danger to the structure. One of the ablest architects in New York makes it a rule to fireproof his columns so that they can be examined at any time by removing the fireproofing to determine whether rust has invaded their capacity to carry their loads. In my judgment, periodical examinations should be made, from time to time, in this way of all wrought iron or steel columns, as it may happen that a leaky steam or water pipe has worked serious harm. Such a discovery was accidently made in an important New York building. *Columns should be stripped*

Numerous newspaper paragraphs appear, at intervals, which claim that metal stripped of its covering of cement has been found exempt from rust, with the paint intact, &c., and the fact is cited as evidence that cement is a preservative of iron and that the danger of rust is overestimated. It is probable that cement will protect *Cement as a preventive of rust* paint for a long time and, of course, paint, if properly put on, will protect iron while the oil in it lasts. Painting, by the way, should be done with the best quality of linseed oil and without the use of turpentine, benzine or dryers. It should be thoroughly applied in three coats, with about a gallon to 400 square feet, but the iron should first be thoroughly cleaned of rust and dirt, by pickling or other process. Paint is rarely properly applied, however, and even when of the best quality, is a preservative of the metal, as already stated, only so long as the oil in it lasts.*

Those who claim to have evidence of the exemption of iron from rust rely, I think it will be found, upon iron which has been under exceptionally favorable conditions, free from dampness, the action of gases, etc., overlooking the fact that a leaking water pipe or steam pipe, or the escape of gases from boiler furnaces, will attack iron and gradually but surely consume it. A notable instance of this is the case of the plate girder of the Washington Bridge over the Boston & Albany Railroad, in Boston, where a quarter inch plate girder was recently found to be entirely consumed in places from the operation of gases from the locomotives passing below.

It is quite common to have advocates of wrought iron cite railroad bridges and the elevated railroad structures of New York as proof of their claims, but if they will take the trouble to

---

*It should not be applied in damp weather but only when the metal surface is perfectly dry.

examine these structures they will discover that, in spite of the fact that they are exposed to view, so that they can be painted frequently, the evidences of rust are unmistakable, especially about the rivets; and one can well imagine what would be the result in the case of riveted iron members in the skeleton structure of a building where such ironwork is entirely concealed from view, periodical inspections being impossible.

Rust is especially liable in the cellars and basements of buildings. The wrought iron friction brakes of freight elevators in the cellars of stores, for example, are frequently found so consumed with rust as to be easily rubbed to pieces in the hand.

STEEL RIVETS are dangerous and they should never be used, unless of a very superior quality, so soft that hammering will not crystalize the material and yet with sufficient tensile strength to insure perfect holding qualities. This is difficult to secure. Their use in columns for buildings is objectionable, as they rust badly under certain conditions. The beam bearing bracket shelf on cast-iron columns, should be cast in one piece with the column and the beams should be bolted to the columns to secure rigidity.

EXPANSION OF IRON.

It is generally supposed and frequently stated that there is a great difference between the expansion of iron and masonry by heat. This is not the case. For example, the length of a bar which at 32 degrees is represented by 1, at 212 degrees would be represented as follows:

Cast Iron,	1.0011
Wrought Iron,	1.0012
Cement,	1.0014
Granite,	1.0007
Marble,	1.0011
Sandstone,	1.0017
Brick	1.0005¼
Fire-brick,	1.0005

In the fireproof building of the Western Union Telegraph Company in New York, some years ago, a heavy brick pier, seven or eight feet in diameter, adjoined the wall of the boiler furnaces. The difference in expansion in the brickwork next to this furnace wall as compared with that of the remaining brickwork of the pier, was so great as to produce a crushing of the material from top to bottom of the pier, for a depth of several inches, and it was found necessary to change the furnace wall and leave an air space between it and the pier.

CONDUCTIVITY.

Expansion While the difference in expansion between masonry and iron incorporated with it is less per running foot than is generally supposed and while the difference in expansion between a cubic foot of iron and that of a cubic foot of masonry would hardly be noticeable, especially if the iron were covered on all four sides; yet in stretches of 50 feet or more, as in the case of iron I-beams and girders, the cumulative effect of expansion in uncovered iron might be a serious matter— quite sufficient, with the rises of temperature due to a burning building, to push out the bearing walls and wreck the building. Especially is this true of temperatures higher than 500 degrees. It is unnecessary to suggest that metal differs from masonry in the important respect that heat does not travel throughout the entire length of the latter, while it does in the case of metal.

In other words, while the difference between the expansion of a lineal foot of iron as compared with a lineal foot of masonry, marble, brick, etc., is very slight, the difference in conductivity is very great. The conducting power of silver, for example, being represented by 1, copper would be .845, cast-iron .359, gold .981, marble .024 and brick .01—an important fact to be considered in the construction of buildings. Brickwork raised to a white heat would not raise the temperature of other masonry in the same wall a few feet away, but one end of an iron I-beam could not be raised to a white heat without raising the temperature of the beam for its entire length.

It is a well-known fact that iron responds so readily to temperature that in surveying land, a surveyor's one hundred foot iron chain will, in measuring the distance of a mile, result in a variation of five feet between winter temperature and summer temperature, resulting in an error of one acre in every 533. Of course atmospheric rises of temperature would not affect the protected structural iron in a building.

PROVISION FOR EXPANSION.

Where iron beams and girders are inserted in walls without sufficient space left for their expansion under heat they are almost certain to overthrow the bearing walls by their expansion thrust. A large warehouse in Vienna in which such provision had been contemplated by the architect was totally destroyed, with its contents, by reason of the fact that an officious subordinate, discovering the space in the wall purposely left at the end of each beam, deliberately poured liquid cement therein, which, having set, effectually thwarted the well meant intention of the architect, and resulted in the destruction of the building.

The expansion thrust of iron beams may be computed upon the following factor of expansion: rolled iron of a length of 1562 feet will expand one-eighth of an inch for every degree of temperature. The heat of a burning building as already stated is enormous—sufficient to fuse most known materials; it may safely be estimated to be at least 1000 degrees; therefore a length of rolled iron of 1562 feet at 1000 degrees of temperature would expand about 125 inches, and a 50-foot length of iron girder would expand between four and five inches, showing that there should be a play at each end of at least two inches if the iron is not fireproofed. Inasmuch as in iron construction the iron beams and girders are usually anchored to the walls to steady them, the space should be left and the tie to the anchor should be by a movable hinge joint, which would be of the same strength with an inflexible anchor for all tying purposes but would yield under the thrust pressure like an elbow and allow play of the beam, or stiff anchors should have elongated holes to allow expansion when beams are of great length. Girders are seldom over 25 feet long, but if bolted together, as is frequently the case, they may be 120 feet or more long, and a line of columns from cellar to roof of a building may easily have one continuous iron structure of two hundred or more feet. It should be remembered, however, that this danger from the expansion of iron may be almost wholly counteracted by protecting it from exposure to fire through the use of non-conducting material. It is more important to protect girders than beams.

The mistaken pride with which the owners of some buildings point to exposed iron beams in ceilings as evidence that the floors are "fireproof," actually justifying the supposition that they are left exposed for such display, would be ludicrous if it were not serious. In buildings occupied for offices or dwellings, where there is not sufficient combustible material to endanger the beams, it is not so objectionable; but in warehouses and stores, filled with merchandise, such construction is dangerous; and if one of the upper floors should give way it would come hammering down to carry all below and thoroughly wreck the structure.

In this connection it is well to say that combustible merchandise should never be stored one hundred feet above the street grade even in a fireproof building, since the average fire department cannot reach it at that height.

Fifth.—The roof, that portion of a building which ought to be most carefully watched during construction, is often the most neglected, woodwork entering into its composition. In the case of the Horne Building, at Pittsburg, the cornice was supported on wooden outriggers.

Roof

Sixth.—PARTITIONS. These should not be erected upon wooden sills, as is sometimes the case— only, however, with ignorant and inexperienced architects, who suppose that it is necessary to use wood in order to nail baseboards and other trim at the bottom of the partition. Porous terra cotta will hold nails and should be used in preference to wood, which as soon as it burns out will let down the entire partition.

Seventh.—All buildings over **75** feet high

should be provided with 4-inch or, better still, 6-inch vertical pipes, with Siamese connections at the street, for the use of the fire department, extending to the roof, with outlets for hose at each story and on the roof. **Water stand pipes** This would save the time of carrying hose to upper floors—a difficult task in the case of high buildings. Ample tanks of water should be provided on the roof supported by protected iron beams resting on iron templates on the brick walls, to supply the building's inside pipe system for fire extinction, and secure pressure by gravity or by some other method constantly operative, especially on holidays and at night. Stone templates should not be used, and care should be taken to secure strong supports so that, in the event of fire below, the tanks will not come crashing through the building to destroy it and endanger the lives of firemen. Two such disasters in fireproof buildings within a year show how true is this proposition. Tanks in the basement under air pressure are also a great advantage and recent invention has perfected them to the point of reliability.

Fire Marshal Swenie, of Chicago, urges that stand-pipes should not be less than six inches internal diameter, and that a check valve should be provided so that, when steamers are attached, their force will be added to that of the local pumps. Each floor should have hose connections with the stand pipes and sufficient hose to reach to the most remote point of the floor above, and this hose should be frequently inspected to see that it is in order. He recommends that a code of signals by which communication can be established between the firemen and the engineer of the building is essential.

Eighth.—All high buildings should have constantly present, night and day, some competent person understanding the elevator machinery, fire appliances, etc., so as to aid the firemen in reaching the upper levels; and there should be sufficient steam in the boilers, at all times, to run one elevator.

<small>Night watchman</small>

I quote from the valuable treatise on handling fires in these buildings presented by Fire Marshal Swenie to the International Association of Fire Engineers held in August 1897. He says:

<small>"In case the elevators fail it is necessary to use the stairway, and after the truck men should follow the pipe men bearing the necessary hose, and this must be carried on the shoulders of the men. A 50-foot section of ordinary 2½-inch cotton hose with couplings weighs from 56 to 60 pounds, and 250 feet of 1¼-inch rope about 65 pounds, either of which is a good load for a man who must climb a steep stairway to the height of 250 feet. With an average rise of seven inches per step, that means taking some 430 vertical steps before reaching the scene of action and consuming from seven to ten minutes of time. If it is found necessary to use hose instead of the standpipes for taking the water from the street to the floor the hose should be taken up in the elevator, if it is running, and then lowered until connection is made with the hose below."</small>

Ninth.—Marble, slate and other stones, are certain to disintegrate or crumble when subjected to the joint action of heat and water. For this reason, ninety per cent of the staircases in modern fireproof buildings would be found utterly unreliable in the event of fire, either for the escape of the inmates or for the use of firemen—a serious consideration. Stone treads are usually let into iron frames, being supported around their edges by a bearing of about half an inch on shallow rabbets in the strings and risers and as the stone treads would give way and fall through in case of fire, it would be impossible for a person to find a footing on the stairways. Two-inch oak treads might actually last longer; but a safer staircase would be one with a frame work of iron, having an iron web or gridiron pattern tread to support the stone or slate, the interstices or openings being small enough to prevent the passage of a foot, so that if the stone tread should disentegrate, the staircase would still remain passable.

It is possible to have the supporting tread of open-work wrought-iron in an ornamental pattern which, in relief against the white marble tread resting on it, would present a tasteful appearance from the underside or soffit of the staircase, with this great advantage that, in the event the action of fire and water should pulverize the marble or

Stone staircase treads

slate tread, it would still afford a safe support for the foot. In the case of the burning of the two fireproof buildings, Temple Court and the Manhattan Savings Bank, in New York, the slate treads yielded early in the fire, leaving staircases with openings the full size of the tread, which, within a few minutes after the fire started, were impassable either for firemen or inmates. It is astounding that this vital fault should be so generally overlooked in fireproof buildings.

I may here state that the Manhattan Savings Bank building did not deserve to be called "fireproof" for the reason that it had hollow spaces under the wooden floor boards and that the iron beams and girders were not protected. Some of them were large, riveted, box girders, which yielded quickly to the heat of burning goods and pushed out the side walls.

It is generally supposed that it is not necessary to be careful as to stone treads in buildings occupied solely for offices separated in fireproof hallways in which, it is claimed, there is nothing to burn; but in the case of one large fireproof building of this kind in New York I found the space under the staircase in the basement story, was used to store the waste paper and rubbish of the building—material particularly likely to cause a fire by concealed matches, oily waste, cigar or cigarette stumps, etc., and to make a lively and quick fire quite sufficient to destroy stone staircase treads. Even where there is no combustible material in the hallway, if the staircase is near windows, stone treads may be destroyed by exposure to burning buildings and by the combustion of window frames, dadoes, wainscotes and other wooden trim.

Tenth.—No building should exceed in height the width of the street on which it is located, from the viewpoint of light and health; nor in any case, in excess of 100 feet for mercantile occupancy, nor a height in excess of 200 feet for office occupancy.

Eleventh.—It should be remembered that merchandise, furniture, etc., are combustible, no matter whether located in fireproof buildings or in ordinary buildings. This obvious fact seems generally to be ignored. **Destructibility of contents** In fact combustible material may sometimes be more effectually and thoroughly destroyed in a fireproof building than in an ordinary building, since the early collapse of the latter may smother the fire and effect salvage, whereas fireproof floors support the contents of the former and distribute them so that they are more certain to be destroyed. There was not a dollar of salvage in the large stock of merchandise in the Horne Building, at Pittsburg. The entire household furniture of a tenant in one of the best fireproof apartment houses in New York was totally cremated; and a fire in the Great Northern "fireproof" Hotel, at Chicago, seriously burned the automatic organ to the extent of over $4,000. There is no more reason why the combustible contents of a fire-proof building should not be consumed than why the fuel in a stove should not be burned.

Twelfth.—ENCLOSING WALLS. These should be of brick, the brick work of the lower stories especially, if not of all, being laid in cement mortar. In fact the specifications for a building in the compact part of the mercantile section of a city ought to be drawn in contemplation of the possible cremation of its contents and the generation of heat considerably greater than 2,000 degrees Fahrenheit. The heat of a wood fire is from 800 to 1140 degrees; charcoal, about 2200; coal, about 2400. Cast-iron will melt at between 1900 and 2800 degrees. Wrought iron 3000° to 3500°; Steel, 2400° to 2600°.

If an architect should be required to draw specifications for a building adjoining others, with the knowledge beforehand that its entire contents, from cellar to roof, were to be totally consumed, and he were under a bond to pay damages to surrounding property, he would not be more severe in his exactions than should a building law protecting neighborhood rights in the enjoyment of property; for a mercantile or manufacturing building sometimes generates a greater heat in combustion than a smelting furnace.

If the architect of a building were actually designing a structure which could safely cremate its own contents, how different would be the nature of the device from the flimsy buildings which form the larger portion of our great cities! When it is borne in mind that where one such building gives way, it follows, as a matter of course—the next on either side being no more substantial than the first—that a conflagration must ensue, the wonder is, not that we have so many sweeping fires in the cities of this country but, that each city is not in turn destroyed, block by block. It is a high tribute to the efficiency of American Firemen that such conflagrations as those of Chicago and Boston are not disasters of annual occurrence, as they would be but for such fire fighters as Bonner of New York and Swenie of Chicago. It is criminal to erect buildings which endanger the valuable lives of such men when a few dollars spent in fireproofing iron work would save them.

Every unsafe building is a death trap for heroic men and the owner who deliberately and knowingly erects one should be held responsible for loss of life. It is safe to assert that every building occupied for mercantile or manufacturing purposes, with naked iron columns, is a fire trap.

Having stated these premises, some of which I have claimed are self-evident, I will proceed to consider, as briefly as possible, without going too much into detail, those features of construction which observation has taught me are most important.

SUMMARY OF IMPORTANT POINTS.

IT is hardly necessary to deal with the foundations of buildings. The question is an engineering problem which does not require suggestions from a fire standpoint, and I shall not deal with it here, other than to touch again upon the important point of not having wrought iron or steel columns in the cellar or basement, where moisture and gas conditions would increase the danger of rust. **Foundation**

ENCLOSING WALLS.

These, as already stated, should be of brick, the lower stories laid in cement mortar, not less than 16 inches thick at the top of the building and increasing 4 inches in thickness for every 25 feet in height to the bottom. This would require a 44-inch wall at the grade for a 200 foot building. The thicknesses here recommended are for buildings not exceeding 100 feet in depth. If they exceed this depth without curtain or cross walls, or proper piers or buttresses, the walls should be increased in thickness four inches for every additional 100 feet in length.

Brick is the best known resistant of fire. Stone yields readily to the combined effect of heat and water, and even terra cotta or burned clay tile cannot be regarded as a perfect substitute for hard burned brick.

Under no circumstances should the iron frame work of a skeleton building be incorporated in thin enclosing walls. No wall that has not a cross section sufficient to support itself without the ironwork, should be allowed, aside from the importance of having it thick enough to prevent the passage of hot air from an adjoining building.

Curtain walls for enclosing walls, supported by the longitudinal members of skeleton construction are objectionable; they are liable to be buckled out by the expansion of the framework. The great trouble with modern fireproof structures, even under the New York Building Law, is that while the separating fireproof floors tend to prevent the passage of flame from one story to another, the enclosing walls are often insufficient to prevent heat from igniting the contents of an adjoining building, so that what is gained by preventing the spread of fire vertically is lost laterally.

It should be borne in mind that the thickness of walls herein recommended is not for carrying capacity as bearing walls. Thinner walls would answer for that purpose. It is intended to confine the heat generated by a fire and should be required in the compact portions of cities, where every man should be compelled to build with reference to the safety of his neighbor.

Architects and builders generally seem to have in mind only the carrying capacity of walls and to lose sight of this important fact.

As the floors and contents of a mercantile building burn they sink to the bottom, where enormously high temperatures are reached, and it is for this reason that walls should increase in thickness as they approach the bottom, on the same principle that the walls of smelting furnaces are thicker at the bottom than at the top. Fireproof floors are not apt to give way, however.

It is the generally accepted opinion that a 12-inch brick wall will prevent the passage of fire, but a much thicker wall may fail to confine the heat of a burning building sufficiently to prevent the ignition of combustible merchandise or other material in an adjoining building. In a fire which occurred in Boston, several years ago, combustible material was ignited through a three-foot wall, which became so hot as to conduct the heat into the adjoining building. In an isolated location an owner might be permitted to construct his walls with reference only to their carrying capacity, but where he builds in the compact part of a city, storing combustible materials from cellar to roof, he should be required so to build that a fire in his premises will not necessarily destroy his neighbor's property. He may well observe a regulation which, in view of the fact that the buildings of his neighbors outnumber his own a thousand to one, will ensure that he will be, in that proportion, the gainer by rules which secure the safety of all though imposed on himself.

I do not believe "skeleton construction" so called should be permitted for stores, warehouses or manufactories in cities, as the walls are not thick enough to confine the heat of burning merchandise.

In some of our western cities, Detroit, Chicago, etc., the practice is growing of using hollow tiling, bonded like ordinary brickwork, 12 inches thick, for enclosing walls, instead of brick, the exposed steel frame being protected by terra cotta slabs about an inch thick. Such a building would burn more quickly than an ordinary wooden joisted building properly constructed. The Leonard Building, in Detroit, destroyed by fire October 7th, 1897, was an example of the great danger of this style of construction. It was ten stories high, and as fast as the columns or wall girders were warped by the heat the tiling dropped out like loose bricks, leaving the entire structure after the fire a ragged cage-work of iron with very little of the tiling on the enclosing walls and none of the floors intact. The contents were, of course, totally destroyed.

PIERS, BOND STONES, ETC.

Bond stones should not be allowed in piers vital to the building or carrying great weights, especially in the cellar or basement.* Stone yields readily and quickly to the combined effects of water and heat and, disintegrating at its edges, gradually releases the bricks above it, so as, in time, to destroy the integrity of the pier. Bond stones are employed by the mason to steady his work. A green brick pier while being laid is frequently unsteady, and a bond stone enables him to progress with his work by steadying all below it so as to receive new courses of brick. In all cases the bond should be a cast-iron plate. If the plate should be cast with holes through it about $1\frac{1}{2}$ inches in diameter, so that the mortar and cement can thoroughly incorporate the plate with the masonry above and below, it would be an improvement. Wrought iron is liable to rust and should not be used. Where bond stones are used in the outer walls of buildings they are less objectionable, but for inside piers they are so dangerous that they ought to be prohibited by law. Strangely enough, only stone for bonds was formerly required by the New York building law, and such was the opposition of the stone men

---

*See Cammeyer Building fire, pages 86 and 87.

to the prohibition of bond stones altogether, (when later it was proposed,) that a compromise was reached allowing the use of cast-iron bonds as an alternative of stone bonds—an option seldom availed of by architects, builders or owners, however, and construed generally by the public to mean that either is good enough.

STONE PILLARS.

It not unfrequently happens that a building of otherwise admirable construction has its weakest point in the cellar, where stone pillars form the support for the entire line of columns through the building. In case of fire and the application of water these stone pillars, no matter how substantial, whether monoliths or stone blocks, will disintegrate and bring down the entire structure. After the great Boston fire, granite piers were shoveled up and carted away like so much sand. It is quite a common practice, and a most dangerous one, to employ single stone columns, often of polished granite, to support the centre of a long stone lintel carrying the wall over an ornamental entrance. Such a column would surely yield to the effect of fire and water and perhaps let down the entire front. In almost every city such faulty architecture may be observed.

The American Exchange Bank, on Broadway, N. Y., recently torn down to give place to a modern, fireproof structure, had a ceiling of carved stone in large squares, supported on the flanges of iron beams. This would have yielded quickly to the heat of fire and water, endangering the lives of firemen or employes. The building was claimed to be fireproof and illustrates prevailing ignorance as to the danger of stone as a building material in a city noted for its knowledge of safe construction.

HOW TO BUILD FIREPROOF. 31

The writer passes every day a costly structure in New York whose corner is supported by a granite monolith column of this kind. If stone columns are desired for architectural effect they should, wherever they carry heavy loads, contain a centre column of cast-iron of sufficient carrying capacity to support the superimposed weight.

CAST-IRON VERTICAL SUPPORTS.

The vertical supports, columns, pillars, etc., as already stated, should be of cast-iron, cylindrical in form, of liberal thickness, especially in the lower stories, thoroughly tested as to sand holes, thin places, from "floating cores," etc. Cast-iron columns should be round, and not square. In the former shape there is less likelihood of defects in casting, sand holes, etc., which prevent uniform sound thickness of shell. The columns should be planed to smooth bearings, so that the entire system from the foundation to the roof, may be securely bolted together and form a continuous line with joints for expansion and without any inequalities of bearings. Under no circumstances should wedges or "shims"* be allowed. This most important matter is often neglected. The flanges and corbel brackets for supporting beams should be cast in one piece with the column and not depend upon rivets or bolts. Rivets, aside from the danger of shearing

---

*"Shims" are pieces of slate or iron inserted to secure a true vertical where the two surfaces have not been properly leveled or planed.

strains, are almost certain to rust to the point of danger. The beams should be bolted to lugs on the columns, however, as a tie between the side walls, holding the entire structure firmly and consistently together as one rigid whole and yet with play for expansion.

Col. Geo. B. Post of New York has devised a form of cast-iron cage construction consisting of pillars and floor beams of the Hodgkinson pattern the members of which lock into each other, without the use of bolts or rivets, forming a very rigid construction and saving the cost of mechanics for bolt and rivet work. While I have not had an opportunity to examine it, I have great faith in his judgment; my impression, from his description of it, is that it would be very rigid construction and admirably adapted to warehouses six and seven stories high. Above this height merchandise should not be stored in any kind of a building.

The factors of safety, in computing strains, should not be less than those prescribed by the standard modern authorities. It is better to be sure than sorry.

All iron work, columns and pillars, beams and girders, should be fireproofed, i. e., covered with at least four inches of incombustible material, terra cotta or brick.

Fireproofing iron members

At the floor, and for a height of four feet in mercantile buildings, a metal guard should be provided to prevent the column from being stripped by collisions with rolling trucks for moving merchandise. It ought to be unnecessary to suggest that wooden lagging should, under no circumstances, be used to protect the iron, were it not for the fact that in one of the largest and most costly dry-goods stores in New York, the fireproof covering of the iron columns, which had been seriously damaged by trucks, was being systematically removed in order to substitute wooden lagging, when the fault was, fortunately, detected by an inspector of the underwriters. Thick hardwood cleats showing the plaster behind might answer as fenders or guards. Four inches of good brickwork is a good covering, but porous terra cotta or even wire lath and plaster may prove effective. Where wire lath and plaster is used the column should first be wrapped with a quarter-inch thickness of asbestos bound with wire. This would prove reliable and inexpensive.

It is a fact, showing how common is the neglect to cover iron with non-conducting material, that in the New York State Capitol, in the library, is a large plate girder entirely exposed. This girder supports the ceiling beams, and there is enough combustible material in the oak bookcases, furniture and flooring to wreck this portion of the building by expansion in case of their combustion. The ceiling of the Senate chamber is of heavy hard wood attached to the soffits of the iron beams, and they would if ignited, probably warp and expand the beams to a dangerous point. The New York Building Law was enacted in this building.

A notable instance showing the necessity of protecting ironwork with incombustible material, and the danger of expansion in long lines of iron girders or beams was that of the destruction of a fireproof spinning mill at Burnley, England, recently. This mill was 210 feet long by 120 feet wide. Six cast-iron girders of the Hodgkinson type, each 20 feet long, spanned the 120 feet width, being bolted to cast-iron columns and carrying, in turn, cross girders of wrought iron. The expansion of these 120-foot girders (they were unprotected) resulted in the disruption of the floor and the destruction of the mill. The cast-iron columns, being unprotected, collapsed under fire and water. The floors were 10' 6" bays. As already stated, beams should not be spaced over five feet on centres. Wider spacing results in weak arches, liable to be buckled out by heat or punched through by the falling of safes or of other heavy articles from upper floors.

The probability is that if the 20-foot girders in this building had been arranged with provision for expansion, and all the ironwork had been thoroughly protected with fireproof material, little damage would have been done. The effect would have been more rapid if the floors had been loaded with combustible merchandise. There was little wood to burn in the contents of the spinning mill, and yet the destruction was thorough. Such buildings with uncovered iron work are more dangerous than those of heavy wood construction, in which the timbers are twelve inches in diameter. A properly constructed building with protected iron, however, is, of course, superior to any other form of building. Experienced firemen are afraid to enter buildings supported by iron columns unless they are thoroughly fireproofed, as they are liable to snap without warning under the influence of fire and water, whereas wooden posts burn slowly and give notice of collapse. They will stand a severe fire without being charred for more than two inches of their surface.

BEAMS AND GIRDERS.

In mercantile buildings and factories, beams, as already stated, ought not to be spaced more than five feet apart, no matter what kind of arch is employed; and while many experts claim that a heavy iron I-beam, thoroughly encased in fireproof material on three sides and having only its soffit or underside exposed, would not be expanded enough by the heat of a fire to cause its collapse, it is best to take no chances but to protect the underside with fireproof material, which can be cheaply applied with wire lath and plaster or by having the skewbacks of the terra cotta floor fillings extend below the soffit or bottom flange of the beam, and made with lips for protecting the iron.

TIE-RODS.

It is a mistake, in my judgment, to dispense with tie-rods, even with the kinds of arches which employ wire cables or other metal ties. The claim is made that these act as tie-rods, but it should be remembered that they cannot be relied on during construction, when derricks for hoisting iron beams and other materials are resting on the girders. Dangerous lateral movements and twistings of the structure may be the result of want of rigidity which, at this stage, can only be secured by tie-rods.

MATERIAL FOR ARCHES BETWEEN BEAMS.

It is my opinion—but there are many who entertain a different one—that the old-fashioned brick arch is the most reliable for resisting fire; that next to this in safety stands the porous, terra cotta, segmental arch, with end construction, i. e., the blocks or separate pieces placed end to end between the beams, instead of side by side in what is known as "side construction." This is said to be stronger than side construction. It is claimed by many experts that porous terra cotta is a better non-conductor than brick on account of its interior air spaces. The arch should not be less than four inches thick, having a rise of at least 1¼ inches to each foot of span between the beams, and there should be a covering of good Portland cement and gravel concrete over this to ensure a waterproof floor. Cinder filling will burn—crushed slag from blast furnaces is better but the Portland cement concrete should not be omitted for water-proofing purposes.

There are many patent floor arches for filling between I-beams which have great merit when properly put in, but I doubt if any of them are equal to the two I have named, and it should always be borne in mind that when employed they should be inserted with the same care with which they are prepared for tests. This is almost equally true, however, as regards brick and burnt clay arches, also. There is less likelihood of poor installation work, however, with brick arches or segmental arches of porous terra cotta or burnt clay. Arches should be laid in cement not lime mortar. They should not be laid in freezing weather, and where concrete is used the broken stone or gravel should be carefully washed and the cement should be of the best quality. Some of the better qualities of patent floors are the following: Fawcett, Guastavino, Rapp (which should be segmental shape) Columbian, Metropolitan, Roebling, Manhattan or Expanded Metal, etc. These floors are fully illustrated in most of the text books on construction. In all of them, I repeat, the spacing of beams should not exceed five feet.

A recent writer says:

"The question of fire-proof material is really a very simple one, and any one who is so disposed can make the most convincing sort of test by taking a small fragment of ordinary porous terra-cotta and a small fragment of the cinders concrete which is usually employed for concrete constructions, and holding a piece of each in his hands, expose the other end to the flame of a blow-pipe. He will drop the piece of concrete first. Some time afterwards he will have to drop the terra cotta. If while hot they are dropped directly into a bucket of water, the most casual inspection will satisfy any one that what is left of the concrete is hardly the material that is most desired for the protection of a building. Concrete is cheap, terra-cotta is not; therein lies the secret of the possibilities of the use of the former material.

Another point. If terra-cotta arch blocks are set in place with only ordinary care, they can be depended upon to serve their purpose. Concrete, on the other hand, has to be mixed most carefully in order to secure a uniform and reliable product. As, in a large building, the bulk of the work is of necessity entrusted to laborers who can be depended upon *not* to think or be careful, the chances are decidedly against a satisfactory mixture of concrete, thereby largely increasing the odds in favor of terra-cotta."

The air spaces in porous terra cotta account for its being so good a non-conductor. They are secured by mixing sawdust with the moist clay before burning, When the material is burned the sawdust is consumed, leaving pores in the baked material.

I do not agree with the writer quoted above in his sweeping condemnation of concrete floors for buildings not over one hundred feet high—and this should be the limit of all structures to contain merchandise, which should never be elevated to a greater height above the grade, owing to the

HOW TO BUILD FIREPROOF.

inability of fire departments to throw water effectively to these heights—I should regard such floors, if properly laid, in non-freezing weather, as good fire-stops for buildings one hundred feet high and under.

WATERPROOF FLOORS.

It is of great importance that the floors of all buildings should be waterproof, in order that the volume of water thrown by the fire department to extinguish a fire may be carried off without injury to merchandise on the floors below. Neglect of these precautions is criminal in view of their simplicity and inexpensiveness.

After the arches have been set between the I-beams they should be covered, for at least a thickness of one inch, with the best Portland cement concrete, carefully laid, so that all water will run to the sides of the building and be carried off by water vents or scuppers, which may be arranged with pipes through the walls having a check-valve which would prevent the influx of cold air and yet admit of the outflow of water.

All ducts for carrying steam, gas and other pipes and electric conduits should be protected with a metal sleeve going above the surface of the floor, and the space between and around the pipes should be filled in closely with mineral wool, asbestos or some other expansive and fireproof material to cut off drafts and flame.

FLOOR SURFACES.

Floor boards should be dispensed with, if possible, (always above 125 feet high) and asphalt or concrete employed instead. It is not popular in office buildings, however, to dispense with wooden floors. Wherever used they should be so laid, especially in mercantile or manufacturing buildings, as to leave no air space to supply a passage for flame and to form a harborage for rats and mice, to which these vermin can carry matches, oily waste or other combustible material, to be ignited by steam pipes or by spontaneous combustion.

FIRE-PROOFING WOOD.

Various processes, "electric," so-called, and otherwise, have been patented for fireproofing wood. They undoubtedly increase the fire-resisting properties of wood for interior trim, window casings, etc. Whether or not they impair the durability of wood is a matter as to which I am not yet informed and I doubt if sufficient time has elapsed for a proper test. The United States Navy has made trials of fireproof woodwork—with what success I am not informed.

VENTILATING AND LIGHT SHAFTS, DUMB-WAITER SHAFTS, ETC.

The enclosures of all ventilating shafts, for water-closets, etc., light-shafts and dumb-waiter shafts should be constructed in the same substantial manner, as freight elevator shafts. It is

HOW TO BUILD FIREPROOF.

a mistake to use thin plaster board or plaster with dove-tailed, or other metal, lath, etc. No enclosure should be relied upon less than four inches in thickness, well braced with angle-iron, but brick walls are best, especially in buildings over 60 feet high. The lights of shafts should be of wire glass, set in metal framework, and ventilators should have metal louvers arranged to secure ventilation but not to increase a draft. Slats should be riveted, not soldered, to metal framework, and the metal framework should flange well over the fireproof material of shaft on both sides. It is possible to finish tin-covered fireproof doors with wooden trim so as to be ornamental, with bead panel-work, etc.

WELL-HOLES.

These should be avoided if the building is to be regarded as fireproof. The Horne Building had one 48'x22'. It is almost impossible to control a fire starting in the lower floors where a well-hole opens through those above. Luxfer prisms are now used to secure light from side windows it is claimed with great success.

A recent fire test of the Luxfer Prism, in Chicago, (March, 1898) is stated to have been satisfactory to Fire Marshal Swenie, as showing that these prisms afford material protection from the heat of a neighboring fire in an exposing building, and that to some extent they are substitutes for iron shutters.

STAIRCASES, ELEVATORS, ETC.

These should be in hallways cut off from the rooms at each story by fire walls and doors, to prevent drafts. It is not so important, and is not so practicable, in the case of office and hotel buildings, as in the case of mercantile and manufacturing buildings; but it is advisable, even in office buildings, to have the staircases, elevators, etc., in a separate hallway, the division walls of which should extend through and above the roof and any skylights should be covered with glass not less than $\frac{1}{4}$-inch thick.

In all buildings where the staircase is not thoroughly and effectually cut off from each floor, with provision for ventilating and carrying off smoke, which might, otherwise, smother persons attempting to escape, there should be two staircases, one at each end of the building. Ten persons are smothered to death for every one actually burned.

SKY-LIGHTS.

It is contended by some that sky-lights should oe of thin glass so that they will break easily and permit the escape of smoke and gas. Smoke is ignitible and, when it accumulates in a building, often spreads the fire from story to story or blows out the walls by the explosion of its gases. But while thin skylights are contended for by many expert firemen, it should be borne in mind that

HOW TO BUILD FIREPROOF. 43

nothing so facilitates the spread of fire as a draft, and it would be better to have the skylights adjusted with appliances for opening them, so that when the firemen arrive on the ground, and not before, they may be adjusted to permit the escape of smoke and allow the firemen to enter the building to see where to work to the best advantage. Unless wire glass is used, a network of wire should be above the glass to guard it against flying embers and another should be suspended beneath the skylights so that when the glass cracks and breaks with the heat it will not injure the firemen below.

ROOFS.

These should be of brick or tile on all high buildings, the roof beams being of iron and, where tanks are supported, of sufficient strength to carry many times the actual probable weight of the water and the containing tank itself. *Roofs*

Slate roofs, on very high buildings, especially on street fronts, are objectionable as, in case of fire, the slates would crack and, falling to the street, injure the firemen. A flat roof of brick-tile is better than any other. *Slate roofs*

All water on roofs from rain or melting snow should be drained from the front or sides to leaders, so as to avoid drip points, from which icicles could be formed. Too little attention is paid to the great danger of injury to pedestrians from falling snow or icicles on very high buildings. This may not be a suggestion strictly germain to this article, but it is a matter so often overlooked as to warrant its being referred to in an article intended to deal more or less thoroughly with the subject of fireproof buildings.

BOILER ROOM.

It is essential to the comfort of those on the first floor of a building that precautions should be taken to cut off the heat from the battery of boilers in the summer season, and this important matter is not infrequently overlooked.

If the boilers are to be in the cellar or basement the ceiling should be high and should be double, with a well defined air space between the two. A ventilator of proper size to carry off the heat accumulating near the top should be provided to the roof.

There are so many admirable non-conducting coverings for heating steam pipes, etc., that it is probably unnecessary to suggest anything on this head. Indeed, the whole subject has little to do with the question of a fireproof building, but may, without impropriety, be touched upon in a treatise intended to prevent oversights of all kinds.

HOW TO BUILD FIREPROOF. 45

ELECTRIC LIGHT INSTALLATION, DYNAMO ROOM, ETC.

The electric light installation of a large fireproof building is an important and complicated matter. To insure safety, reference should be had to the rules of the National Board of Fire Underwriters, which can be obtained, without charge, from the nearest local board of underwriters.

The switchboard should be of incombustible material, and no steam, water or sprinkler pipes should pass over or near it where, in case of a bursting pipe, water could reach the switchboard and cause disaster. This is an important matter almost universally overlooked.

An admirable floor for a dynamo-room is one of deck-glass, $\frac{3}{4}$ inches thick, on a wooden (not iron) frame. It will insure that the attendant upon the dynamos will be, at all times, effectually insulated. Such a floor will not become soaked with oil, as would a wooden floor, and can easily be kept clean. A strip of rubber floor carpet stretched over it will prevent slipping. The Continental Insurance Company has, probably, the only floor of this kind in the country, in its large fireproof office building on Cedar Street, New York.

CHIMNEYS AND FIREPLACES.

Nothing tends more to the discomfort of the occupants of a building than smoky fireplaces, whereas, on the contrary, an open fireplace, with a good draft, is a great comfort.

Unfortunately, the subject of fireplaces and chimneys is frequently neglected by those architects who pay more attention to artistic effects than to practical provisions which insure comfort.

The accompanying diagram illustrates how a fireplace and flue should be constructed to insure a good draft. It will be observed that the back of the fireplace is inclined forward, commencing at a point six courses of brick from the bottom, so as to contract the throat of the flue; a square shelf being left, against which downward currents of smoke would strike and rebound, to return up the flue. It is possible and sometimes necessary to have a cast-iron plate resting on this shelf, which can be drawn forward, as occasion may require, to contract the throat of the flue, the capacity of which should correspond somewhat with the size of the fireplace opening, to the extent of having an area about one-eleventh of the latter.

Flues for fireplaces burning wood should not be less than 8"x12" and should be lined with flue tile, which will insure a smooth flue and also tend to the safety of the building. The height of the fireplace should, as a rule, not exceed 25 inches, and the front opening should be supported by two iron bars $\frac{1}{2}"$x2," nine inches longer than the width of the opening.

HOW TO BUILD A CHIMNEY.

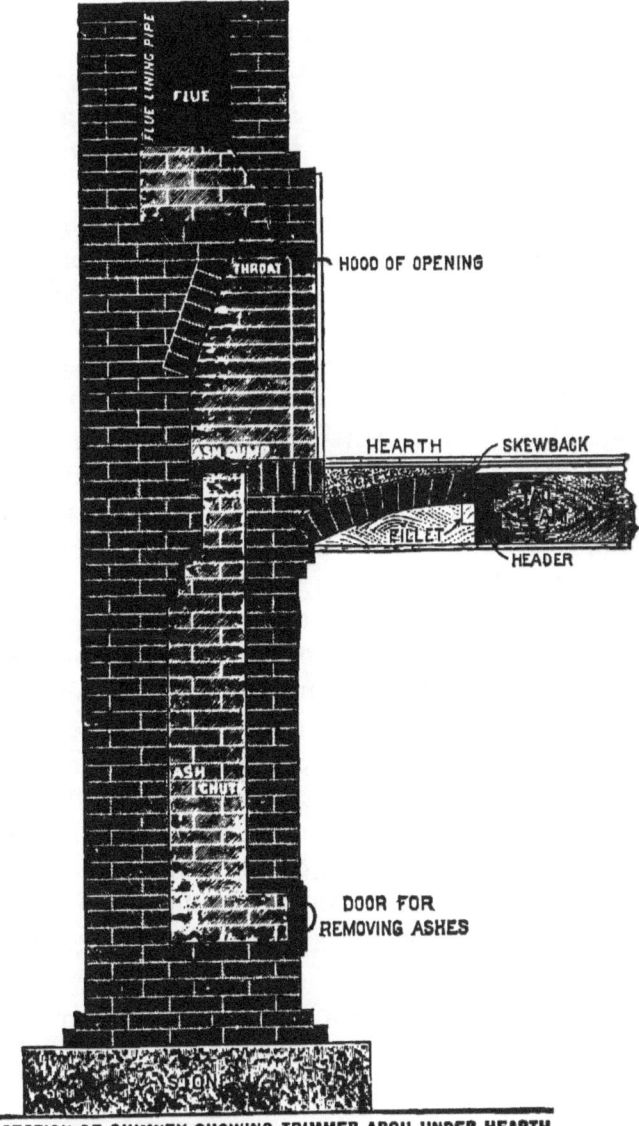

SECTION OF CHIMNEY SHOWING TRIMMER ARCH UNDER HEARTH, PROPER CONSTRUCTION OF FIRE PLACE, FLUE AND FLUE LINING, ASH CHUTE, ETC.

COMMUNICATIONS BETWEEN ADJOINING BUILDINGS.

It is sometimes necessary to have communications between adjoining buildings by doors in the fire walls and it is not always convenient, when changing merchandise from one room to another, to have fireproof doors closed during working hours. It is possible to have the fireproof doors run upon trolleys on an inclined track so as to close by the force of gravity and held open by fusible metal latches or links which would release them when melted by the rising temperature of a fire. It has occurred to me that this difficulty may, also, be met by erecting between two adjoining buildings a separating fireproof hallway of brick, which can be utilized for containing staircases and elevators and for supporting the water tanks of automatic sprinklers. The doors which open into this hallway should not be opposite each other, but at opposite ends of it, so that fire in one of the buildings passing through the door would come against a blank wall opposite. Even if the fireproof doors to these openings should happen to be open at the time of a fire in one of the two buildings, it is improbable that it would find access to the other.

The floors should be both fire and waterproof, slightly lower than those of the two separated buildings, and with water vents or "scuppers"

HOW TO BUILD FIREPROOF.

for carrying off surplus water thrown by a fire department. Indeed it is well to have "scuppers" on all floors of every building.

The walls of this separating hallway or vestibule should rise four feet higher than the roofs of the two buildings and, if there are window or door openings near it, its walls should project beyond the line of enclosing walls at least one foot.

It ought to be unnecessary to state that there should be no combustible material whatever in this separating hallway, and that the staircase, elevators, etc., should be of metal and fireproof.

Indeed such a hallway as this could be relied upon to separate wooden buildings. It should, however, for that purpose, be at least ten feet higher than the peak of their roofs and should extend four feet beyond their front and rear lines. It is probable that the extensive frame dairy buildings of ex-vice-president Morton at Ellerslie, which burned several years ago, might have been saved by this simple precaution.

Separation of wooden buildings

The following diagrams fully illustrate the idea.

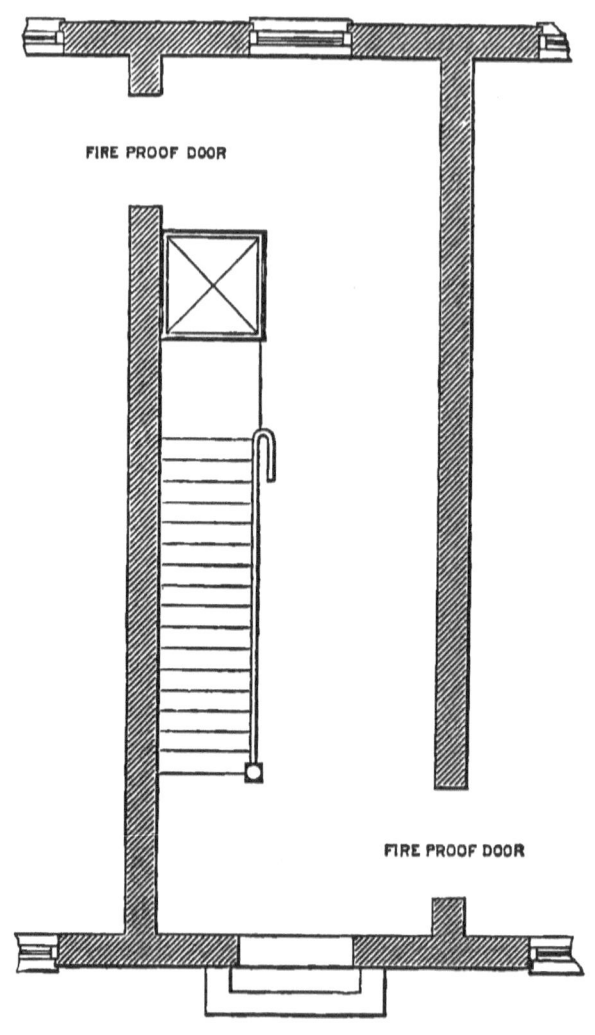

SEPARATING FIRE-STOP HALLWAY, GROUND PLAN.
Scale ⅛ inch to foot.

SEPARATING FIRE-STOP HALLWAY—ELEVATION OF UPPER STORY AND ROOF WITH WATER TANK.
Scale ⅛ inch to foot.

Where it is not necessary to transfer merchandise from one building to another, but only requisite to have a passageway for employees, this may be arranged by an iron balcony, like a fire escape, cutting down the window on each side of the separating wall for a door, so that communication can be had by the balcony. The openings should have fireproof doors. This would be practically safe. It might, with iron ladders, be utilized as a fire escape, and so prove of great advantage to firemen in fighting a fire, who could hold a hose nozzle at the different windows with perfect safety to the last moment. It is practicable, indeed, to have iron stairways with roofed balconies entirely outside of storage stores so that the floors do not communicate. There is a number of these in Philadelphia.

<small>Outside staircases</small>

WATER TANK.

The water tank, as already stated, should be supported on protected iron I-beams, resting on the brick walls, with cast-iron templates, so that the tank cannot fall, break down the staircases and wreck the building in case of fire.

It is important always to locate tanks so that they will not be over stairways or elevators and endanger them in case the supports give way. With a fireproof hallway of the kind recommended, containing no combustible material whatever, the tanks being supported by iron I-beams resting on the brick walls, this would not be an important matter, but in all other cases water tanks should be planned so as not to endanger staircases, and the supporting iron beams should be fireproofed, that is, covered with fireproof material.

FIREPROOF DOORS AND SHUTTERS.

These should not be of iron, but of wood covered with tin. Solid iron shutters or doors are not reliable. Iron doors yield readily to flame, resulting sometimes in their warping open when exposed to fire in an adjoining building, exposing the one they are intended to protect to the full effect of the flames.

Where window openings are protected by iron shutters on rear courts they are almost certain to be warped open by fire in exposing buildings and cannot be relied upon. The tin covered wood shutters alone are reliable. There is no recorded instance in which a solid iron door exposed to the full effect of fire in an adjoining building has protected the opening, whereas there is, on the other hand, no recorded instance in which the "Underwriter's" door has failed to serve its purpose—two important facts which are significant and ought to settle the question.

The "Underwriter's" door is constructed of ordinary white pine lumber, free from knots, of double or treble thickness, according to width of opening, the boards being nailed diagonally and covered with the best quality of tin, with lap-welded joints.*

It ought to be unnecessary to state that, on the exposed side of a building, not only the shutter but the window frame, sash, etc., should be of metal or covered with metal—riveted not soldered. Where it is not possible to use a fireproof shutter for want of room, wire glass in a metal frame will be found a desirable substitute. It will probably hold a fire until the fire department can cope with it. †

It is not generally understood or known that fire will travel from one story to others above by way of the windows in the outer or enclosing walls. Especially where a building has an enclosed court fire will sometimes reach upper stories in this way, even when the floors themselves are thoroughly cut off; the court acting as a chimney. This happened several years ago in the Temple Court Building, a fireproof structure in New York. The woodwork on several floors was ignited by the lapping of fire through the windows from the lower stories and serious damage resulted. A more recent instance was the Livingston fireproof building in New York in

---

*Full specifications for these are given herewith, being the rules of the New York Board of Fire Underwriters.

†Where the exposing building is within 15 feet the glass should be double thickness with an air space between of one inch.

January 1898. Windows on exposed sides should always be protected with fire-resisting shutters.

It may be well to suggest for the benefit of those who are not familiar with city fires that, as heat naturally ascends, the exposure of a low building is often much greater to a neighbor higher than itself than to a building of its own height, so that a tall fireproof structure surrounded by smaller buildings should be provided with fire shutters to all openings. These are not necessary where the exposing buildings are occupied for offices and are themselves fireproof, as the amount of heat which escapes from the windows of a burning building so long as its enclosing walls remain intact is seldom sufficient to ignite a fireproof building or its contents. The moment of greatest danger is when a burning building collapses and the intense heat caused by its enormous bed of coals exerts it full effect upon surrounding structures. In a recent fire in New York (February 11, 1898,) three fireproof office buildings were more or less damaged with their contents although many feet away from the burning building.

It is to be hoped that some inventive genius will devise a plan for simultaneously opening or closing the shutters on any or all stories of high buildings by manipulation from the ground floor. They are usually left open at night—always in the daytime—and might thus be closed in case of a dangerous fire in the vicinity. In some cases they will actually be found fastened open.

Fireproof shutters might be arranged to run on trolleys and close themselves automatically, by force of gravity, when released by fusible links or triggers, which, while unlatching under the heat of a neighboring fire, might also be released mechanically from the ground either by screw rods to lift the triggers or by levers. They should be about three inches wider all around than the openings to be covered, and when not needed could be adjusted to rest between the windows against the piers of the building.

COMPARATIVE TESTS OF FIREPROOF MATERIAL.

Tests of fireproof material, iron beams, pillars, floor arches, etc., to be of any value must be conducted under circumstances which insure uniform conditions, otherwise comparisons are unreliable. It is quite customary to refer to results of fires in different buildings, having differing forms of construction, as supporting theories of relative merit; but ordinary conflagrations cannot be relied upon, for the reason that in two buildings, side by side, the conditions may be widely different. Eddies and currents of air, changes of prevailing wind, etc., may secure exemption from damage. It happened in the large conflagrations of Chicago, Troy, Boston, etc., that the most phenomenal escapes were observed.

HOW TO BUILD FIREPROOF. 57

In some instances frame buildings, surrounded by brick structures which were totally destroyed, escaped with no further damage than the blistering of paint.

Even where tests are carefully arranged, especially weight tests, obvious precautions are sometimes overlooked. It will be observed, for instance, where bricks are piled on a surface of floor arch and iron beams to secure a certain weight per square foot, the pile of bricks may be so disposed as to have a bearing on both of the iron beams and the full weight may not come upon the fireproof arch between them. The lateral bond of a pile of bricks a few courses higher than the floor to be tested, may have all the effect of a relieving arch and materially reduce the strains. In furnaces constructed to secure high temperatures, drafts and currents of air should be provided for with great care and under the direction of the most competent and intelligent experts.

THE INTEREST OF UNDERWRITERS IN FIREPROOF CONSTRUCTION.

In conclusion it may be well to state, in view of the general misapprehension which prevails with regard to the interest of the fire underwriter in the improvement of construction, that it makes no difference to him whether a building be fireproof or not; his rate of premium and the amount which he insures are both based upon the characteristics of each building insured. He would make just as much money on $100 of premium secured at a rate of 5% (or $50 per $1,000) for $2,000 insurance on a wooden planing-mill, as on $100 of premium secured on $100,000 insurance on a fireproof building the rate of which is $1 per $1,000.

RECAPITULATION.

In order to save anyone who contemplates erecting a fireproof building the trouble of revising the preceding pages, I have prepared the following recapitulation of important points to be observed, in order that he may be able to check off his plans and specifications and see that all important features have been duly attended to.

ENCLOSING WALLS. Should be not less than 16 inches thick for the top story, increasing four inches in thickness for every 25 feet to the bottom. Should be built of hard-burned brick, the lower stories (if not all) laid in cement mortar.

All weight carrying walls should be separated by air spaces from furnace walls.

All templates should be of cast-iron, especially for beams which support tanks. Stone templates should not be used.

IRON MEMBERS. All ironwork should be fireproofed *i. e.* protected by not less than 4 inches of fireproof material. Brick is best, well-burned terra cotta second, metallic lathing and plaster third. If plaster or metal lath be relied on, wrap the column with asbestos quarter inch thick, bound with wire. If mercantile or manufacturing building, protect the fireproofing material of the lower four feet of columns with a metal cover, to prevent its being knocked off by roller trucks and merchandise. Heavy hardwood cleats may secure this.

Columns should be cast-iron, the beam bearing corbel brackets being cast in one piece with the column. Columns should be cylindrical (not square) to secure more perfect castings. See that top and bottom bearings are planed smooth and true; no wedges or shims allowed.

Allow for expansion in long systems of beams or girders. Avoid steel rivets; all rivet-work dangerous on account of rust. Beams should be bolted to lugs on cast-iron columns.

All ironwork should be well painted with good linseed oil paint, the iron being first thoroughly cleaned. Avoid turpentine, dryers, &c.

See that fireproofing is applied so that columns may be stripped and examined from time to time.

BEAMS should not be spaced wider apart than five feet on centres.

BOND STONES.—Avoid in piers.

STONE COLUMNS.—Avoid.

TIE RODS.—Do not omit them.

FLOOR ARCHES.—Best, old-fashioned brick arch; next best, terra cotta segmental arches, end construction. If patent concrete arches used, be careful to see that good quality of cement is employed and the stone or gravel thoroughly washed. Arches should not be laid in freezing weather. Only cement mortar should be used and every square foot carefully watched in process.

Cover top with cement concrete to insure waterproof floors. Leave scuppers or water vents at each floor to carry off water thrown by fire department.

Leave no hollow spaces below wooden floor boards.

STAIRWAYS, ELEVATORS, DUMBWAITERS, ETC. Should be cut off in all buildings by a brick partition between the hallways and main rooms, with fireproof doors (for which see Underwriters' specifications.) It is best to have all stairways enclosed in brick walls.

Avoid stone treads, slate or marble, unless web support of iron beneath. It is claimed wrought-iron support is better than cast-iron open work.

Thoroughly fire stop all openings for gas, steam pipes or electric wires, to prevent fire traveling from story to story. These should be in staircase tower.

GLASS WINDOWS.—If on exposed sides protect with fireproof shutters, Underwriters' specifications. Set eyebolts for hinges when building walls. If wire glass be used, they should be glazed in metal frames, and if on exposed side, should have double sheets with one inch space between them.

DYNAMO ROOM.—Avoid water or steam pipes over switchboard. Have glass floor.

FIRE EXTINGUISHING APPLIANCES.—Have 6-inch standpipes with outlets for hose at each story for use of firemen, Siamese connection at street. Arrange signals to street and hose on each floor to reach most remote point.

Have pressure tanks in basement and support all roof tanks on iron beams (fireproof) resting on cast-iron templates on brick walls where they cannot endanger staircases.

Vertical pipes should be in staircase tower.

ROOF.—Avoid all woodwork in roof, even outriggers for cornice. Avoid slates on slanting roofs, as in falling they would injure firemen. Best roof is flat brick or tile.

PARTITIONS must not rest on wooden sills or bases.

NIGHT WATCHMAN.—Have some one on premises at night and on holidays understanding elevator, force pumps, etc. Have enough steam up at all times, to run one elevator.

SKYLIGHTS.—Protect with wire netting above and below and arrange so as to be opened by firemen for letting out smoke and gas. If wire glass is used then no overnetting or undernetting will be required.

CUT OFFS AT STREET for Gas and Electric installations should be provided where firemen can see and use them in case of fire. This is an important matter.

NON-FIREPROOF BUILDINGS.

HOW TO BUILD A SLOW-BURNING STORE OR WAREHOUSE BUILDING
IN ORDER TO SECURE A LOW INSURANCE RATE.

This question is frequently propounded to underwriters by property-owners who, while unwilling to go to the expense of fireproof construction, are quite willing to follow methods which would insure slow combustion, provided the expense of building be not materially increased; especially if lower insurance rates would yield a reasonable return on the expenditure from an investment viewpoint.

All systems for fixing rates of premium for fire insurance by schedules, and especially the Universal Mercantile Schedule, recognize the various features of a building by a system which charges in the rate for departures from proper methods of construction; and especially for those features which tend to retard the progress of fire once started throughout the various stories of a structure and those which aid in its extinction. It is important, therefore, not only that the enclosing walls should be of sufficient thickness to carry the superimposed weight, but that they should be of sufficient thickness to exclude the heat of an outside fire; and it is also important that the various floors of a building should be sufficiently fire-resisting to prevent the spread of fire from one story to another. For this reason, openings through the floors for stairways, elevators, dumb-waiter shafts, channels for water, gas or other pipes, etc., etc., should be cut off at each story. There should, as a rule, be no hollow spaces where a fire could gain headway unobserved, such as spaces between the plastering of a ceiling and the floor boards of the story above; cocklofts or spaces in unoccupied attics; hollow wooden box cornices, etc. Concealed spaces are objectionable, also, for another reason, viz., that they become harboring places for rats, mice and other vermin, to which they carry rubbish, oily combustible material, friction matches, etc., for their nests, which are not infrequently constructed in proximity to flues and heating pipes whose warmth attracts them.

In fact, the chief aim of the architect, from a fire-resisting standpoint, should be, first, to avoid conditions which would favor the starting of fires and, second, to observe precautions which would prevent their spread and facilitate their extinguishment.

The use of wood for beams, girders and supporting columns is not so objectionable from the fire standpoint as is generally supposed, provided they are of sufficient size to carry their loads after their surface has been invaded by fire to the extent of say two inches. Fire seldom gets deeper into a solid 12 inch column or beam with a good fire department. In slow-burning construction or mill construction, so called, all wooden beams, girders and pillars should be not less than 12 inches thick, and the floor plank should be not less than three inches in thickness, tongued and grooved, or connected by splines, with a floor board one inch thick and planed. If Salamander or other fire retardent is used between the two, the probabilities of confining a fire are increased. Sheet iron or tin, painted on both sides with a good oil paint, inserted between the two would be an admirable precaution, but would add to the expense beyond the figure which most property-owners would approve.

The following is the description employed in the Universal Mercantile Schedule for a standard building, to secure the basis rate:

NON-FIREPROOF BUILDINGS.

A Standard Building is one having walls of brick or stone (brick preferred,) not less than twelve inches thick at top story (16 inches if stone,) extending through and 36 inches above roof in parapet and coped, and increasing four inches in thickness for each story below to the ground—the increased thickness of each story to be utilized for beam ledges. Floors of two inch plank, (three inches better) covered by ⅞ or one inch flooring, crossing diagonally, with waterproof paper or approved fire-resisting material between (if tin or sheet-iron between, see deductions;) wooden beams, girders and wooden story posts or pillars twelve inches thick, or protected iron columns; elevators, stairways, etc., cut off by brick walls or by plaster on metallic studs and lathing; communications at each floor protected with approved tin-covered doors and fire-proof sills; windows and doors on exposed sides protected by approved tin-covered doors and shutters; walls of flues not less than eight inches in thickness, to be lined with fire-brick, well-burned clay or cast-iron, and throat capacity not less than 64 square inches if steam boilers are used; (12 inch circular flues are better) all floor timbers to be trimmed at least four inches from outside of flue; heated by steam, lighted by gas; cornices of incombustible material; roof of metal or tile; if partitions are hollow or walls are furred off there must be fire-stops at each floor.

If this description of building is still further improved, reductions in rate may be secured for the following features of exceptional construction:

FLOORS. For tin or sheet-iron between floors.

If waterproof, arranged with waste-ways and scuppers and inclined to carry off surplus water thrown by fire department to sewer or street.

If the grade floor be fire-proof, protecting the upper portion of the building from fires in the basement or cellar, and the communicating stairway to main building thoroughly cut off.

CEILINGS AND PARTITIONS. If of incombustible material throughout or plastered on metallic studs and lathing, the lathing to have a good key for plaster a further deduction is made.

NOTE.—If the main "fore and aft" partition separating halls containing stairs and elevators from the stores, be of brick or metallic lathing, etc., with protected openings it would save building from charges 67, 68, 69, 70, etc., as to elevators and stairways.

Taking the building from the foundation to the roof, therefore, in the order in which it is constructed, the following details of construction should be observed to secure the lowest rate of fire insurance:

FOUNDATION. This is largely a question of engineering, but it may be stated here that the most competent experts in engineering, architecture and construction, to-day, pay great attention to secure footings and substantial foundations, driving piles to a solid bearing wherever necessary.

STAIRWAYS, ELEVATORS, DUMB-WAITERS, channels for pipes, etc., should be cut off at each floor and enclosed with fire-proof materials; the stairways and elevators especially by brick walls or by fire-proof terra cotta not less than four inches thick, securely braced with angle iron. Brick walls are decidedly preferable. Doors entering from halls to the various rooms should be self-closing and are improved greatly by being covered with tin.

FLOORS. These, as already stated, should be solid, without air spaces, with 3-inch plank, splined or tongued and grooved, and inch thick floor boards; Salamander or waterproof paper between. Tin or sheet iron painted both sides is better.

STORY POSTS, BEAMS, GIRDERS, ETC. These should be not less than 12 inches in diameter, if round, or 12 inches square. The floor beams should be cut on a bevel of three inches where they are inserted into the enclosing or bearing walls, so that, in case of burning through in the middle, they would release themselves without tearing out the walls. There are some excellent patent devices for anchoring floor beams, consisting of cast-iron boxes resting in the wall so constructed as to release the floor beam without damage in case it should burn, and serving also to protect the ends of the beams from dry rot and from charring in case of a fire in an adjoining building.

FLUES. These should be surrounded by at least 8 inches of brickwork, and will be improved still further by having fireproof tile linings.

ENCLOSING WALLS. These should be not less than 12 inches thick for the top story, if of brick (16 inches would be better,) and should increase in thickness four inches for each story to the bottom. While the wall here recommended, and the standard of the Universal Mercantile Schedule already quoted, 12 inches thick at the highest point, increasing four inches for each story to the bottom, utilizing the increased thickness at each story as beam-bearing ledges, is unquestionably the only kind that should be erected for fire-resisting purposes, it is customary to build according to the New York Building Law, the requirements of which are as follows:

"The walls of all warehouses, stores, factories and stables, twenty-five feet or less in width between walls—

Shall not be less than twelve inches thick to the height of forty feet.

NON-FIREPROOF BUILDINGS.

If over forty feet in height, and not over sixty feet in height, the walls shall not be less than sixteen inches thick to the height of forty feet, or to the nearest tier of beams to that height, and from thence not less than twelve inches thick to the top.

If over sixty feet in height, and not over seventy-five feet in height, the walls shall not be less than twenty inches thick to the height of twenty-five feet or to the nearest tier of beams to that height, and from thence not less than sixteen inches thick to the top.

If over seventy-five feet in height, and not over eighty-five feet in height, the walls shall not be less than twenty-four inches thick to the height of twenty feet, or to the nearest tier of beams to that height; thence not less than twenty inches thick to the height of sixty feet, or to the nearest tier of beams to that height, and thence not less than sixteen inches thick to the top.

If over eighty-five feet in height, and not over one hundred feet in height, the walls shall not be less than twenty-eight inches thick to the height of twenty-five feet, or to the nearest tier of beams to that height; thence not less than twenty-four inches thick to the height of fifty feet or to the nearest tier of beams to that height; thence not less than twenty inches thick to the height of seventy-five feet, or to the nearest tier of beams to that height, and thence not less than sixteen inches thick to the top."

No building for the storage of merchandise should be higher than 60 feet from the ground, unless fireproof throughout, and then not over 95 feet, even in cities with good fire departments.

PARAPET WALLS. The enclosing walls of a building should be carried above the roof and coped with stone as a protection from the weather, to a height of at least 12 inches; to protect the building from fires in adjoining structures.

ROOF. This should be of metal, without an air space.

WATER TANK. This should be supported upon brick walls, so as not to give way and fall, as would be the case if wooden supports were consumed. It may rest upon railroad iron or I-beams carried from wall to wall. Under no circumstances should it rest above the staircase, where in falling it would endanger the lives of firemen.

ELECTRIC WIRING. This should be installed in accordance with the rules of the National Board of Fire Underwriters, which may be obtained without charge from any local board of underwriters.

WATERPROOF FLOORS. Great damage usually results to stocks of merchandise from the water thrown by fire departments to extinguish fires. The floors should, therefore, be waterproof and should be so inclined to the side or rear walls that the water will run off by means of scuppers or metal pipes inserted in the walls at the floor level, having check valves which would prevent the ingress of cold air and permit the egress of water. The door-sills should be one inch high. Such precautions are always recognized in rates by underwriters, and it may happen that a fire can be extinguished on one floor without having the water escape to those below.

CLOSETS. There should be no closets, especially for oils, filling lamps or other purposes, under the staircases or elevators, where a fire starting would quickly reach the floors above. In fact closets are always objectionable in mercantile and manufacturing buildings. They are hiding places for careless employees, who put greasy overalls, oily waste or other rubbish in them, often with friction matches. Numerous fires start in such places and assume dimensions under conditions that make them dangerous. Wherever practicable, concealed places of all kinds should be avoided.

RECEPTACLES FOR WASTE, RUBBISH, ETC., These should always be of metal—never of wood—and they should be emptied every night, not left in the building. An old-fashioned cast-iron kettle, with legs and a metal cover, is the best receptacle for oily waste or rubbish, which is liable to ignite spontaneously. All rubbish should be treated as dangerous.

Saw-dust spittoons should not be allowed under any circumstances. They are receptacles for cast away cigar-stumps and cigarettes.

HEATING. If by steam pipes they should, at no point, come in contact with wood, but should be guarded by thimbles where they pass through floors. If by furnace, the hot air pipes should not pass between the floor of one room and the ceiling below, or between stud and lath and plaster partitions. Where it is necessary to have a hot air pipe pass out of sight it should be double with an inner and outer pipe and a space of half an inch between the two.

DRY ROT. It is important to observe precautions to insure against dry rot in buildings with wooden floor joists or wooden columns, especially if they are covered up by plaster to protect them from fire. It is customary to cover them with wire lathing and plaster, and in such cases, small perforations about ¼-inch in diameter through the plaster at the top and bottom of a pillar or column would probably secure sufficient ventilation to save the column, which, also, should be centre-bored. This is true also of floor beams, which may be ventilated at each end with small holes in some ornamental pattern. It is not generally known, however, that unprotected beams, if 12 inches in diameter, as already stated, are rarely consumed to the point of breaking, if the city has even an average fire department.

WOODEN CEILING AND SHEATHING ON SIDE WALLS. This is decidedly dangerous, especially where pine or other resinous wood is used. Fire flashes readily over the entire surface and quickly gets under such headway as to defy the efforts of the fire department. If floor joists and side walls, are not to be left exposed, (the plastering, if any, on the side walls without wooden laths, furring, etc., what is known as "open finish,") the old fashioned plaster, even on wooden lathing, is infinitely preferable to wooden sheathing or ceiling. If plaster is used it should always be upon wire lathing. This insures a good key or clinch to the plaster and will retain the plaster when saturated with water longer than wooden laths. It forms an effective fire-stop for a considerable time and materially aids the fire department in extinguishing a fire.

FIRE-EXTINGUISHING APPLIANCES. Standpipes, supplied by the tank on the roof with water under pressure, not less than four inches in diameter, are always recognized by underwriters. They should have hose outlets at each story, conveniently located near the stairs, where they can be used to the last moment, and the hose should be frequently tested, thoroughly dried after using and arranged so that it will not be rotted by water left in it. Fire pails are admirable for extinguishing fires, and a sufficient number filled with water, with possibly a cask for auxiliary supply, should be near the stairs on every story. They are always recognized by underwriters and very favorably regarded, since the most ignorant persons know how to use them. Where inflammable oils are kept it would be an improvement to have one or two pails filled with sand. Not less than six fire pails to every 2,500 square feet of floor area should be provided. Salt in the water may prevent its freezing in winter.

SPRINKLER PIPES. The piping for automatic sprinkler pipes should be installed when the building is being erected. The rules, sizes of pipes, etc., can be obtained of the underwriters.

WELL-HOLES, HATCHWAYS, ETC., These are always objectionable, as they insure the rapid progress of fire throughout the various floors into which such well holes open. They should be avoided in store and warehouse buildings.

SKY-LIGHTS. These should be of thick glass, with metal frames, and should be guarded by wire netting above, to protect them from falling fire brands from outside fires, and by wire netting below, to prevent broken glass falling on the firemen when extinguishing a fire.

It will be observed in reading the foregoing specifications that, from a fire standpoint, the aim of the architect or builder should be to secure substantial enclosing walls; substantial floor supports; fire-proof enclosures for elevators, staircases, dumb-waiters and all communications from story to story; to avoid hollow concealed spaces; and to secure a fire-resisting roof. These are the main points to be kept in mind, and the necessary details and precautions to be observed ought naturally to occur to any intelligent and conscientious architect or builder.

Perhaps the four most important considerations to be observed in slow-burning construction are:
Timbers not less than 12 inches thick;
Floor planking double and four inches thick and waterproof;
Openings from story to story cut off, to prevent drafts, and
Entire absence of concealed spaces which would afford a harboring place for rats and mice or admit of a fire getting beyond control before discovery.

FIRE IN THE HOME LIFE BUILDING,
Broadway, New York, Dec. 4, 1898.

This building was 55 feet front by 101 feet deep, fronting east, on City Hall Park, with two square light courts, one on the north side and one on the south. It was 16 stories high above the sidewalk; 192 feet to the flat roof, and about 50 feet additional to the top of the tower; constructed with steel skeleton framework covered with incombustible material; arches of hard-burned clay tile; windows on all sides unprotected by iron shutters; window-frames and sashes of wood. A ground plan and photographs of the building taken before and after the fire are herewith shown. This building was in many respects exceptionally well built with iron work protected and stone ashlar well backed with brick.

Fire gained access from the windows in the north court, blown with blow-pipe intensity by a gale of about fifty miles an hour from the northeast. The Fire Department was unable to get water above the eighth floor, or about 108 feet above the sidewalk, and the contents of the stories above this height were almost entirely cremated—office furniture, floor boards, doors, window-sash and frames and other so-called wooden "trim." No damage was done to the iron and steel structure, which was well built, but the marble front, elaborately carved, with projecting balconies, etc., was almost entirely ruined above the eighth floor by the joint action of heat and water.

If the windows on the north side next the exposing building had been protected with fireproof shutters it is probable that little if any damage would have been done in the Home Life Building.

It is well to note that the iron lattice girders or braces which spanned the north court and which may be seen in the photographs expanded and sagged in the intense heat of the adjoining fire and had to be replaced. If these had been boxed with metal, even with tin or possibly copper casing with an air space they would not have yielded. A commendable feature of this court was the use of iron for lintels instead of stone.

The following are the lessons of the fire, emphasizing the suggestions to which reference has been made in the preceding pages.

FLOORS. These, if of wood, should have the space between the underside of the floor boards and the top of the fireproof arches thoroughly filled in with concrete (see page 40.) In the Home Life Building this was not done, and the floors were thoroughly consumed, adding to the destruction of the furniture, etc., and helping to spread the fire from room to room. The spaces under the door sills should always be firestopped, even if the entire hollow space beneath the floor boards are not filled in. The fireproof partitions were, very properly, carried through the floors to the fireproof arches, and did not rest on the wooden floor boards or the wooden sills, as is sometimes the case (see page 16;) but this precaution did not save the structure, because of the openings through partitions for windows, etc.

Another serious fault of hollow spaces under the wooden floor boards was developed in this fire. The floors naturally burned more rapidly under the front of fireproof safes and let down the front castors or wheels in advance of the rear wheels. The falling forward of a large safe on one floor broke out the fireproof arch and let the safe through to the floor below, where fortunately it was arrested and held by a floor beam. Fireproof safes should be arranged so as to extend over two of the iron floor beams, or if the spacing of these is too great they should rest on iron bars or plates so as to extend the bearing to the beams. This is important in the case of jewelry safes and would be a serious feature in arches of wider spans than five feet—the Guastavino arches for example.

There can be no question that all buildings in excess of 125 feet high, or certainly those portions of them above this height, should have fireproof floors of concrete. Such floors can be easily arranged with holes at the border for fastening floor cloths, carpets, etc. Only cement, rock asphalt, marble tiling, mosaic or other incombustible materials should be employed.

WINDOWS. All window frames or sash should be of metal and protected with fireproof shutters (page 6.) Wherever shutters are not possible, wire glass should be used.

STAIRCASES. Stone treads of marble or slate should not be used unless supported by an iron web of sufficiently close meshes to afford a footing in case the stone yields to fire and water, as it certainly will if exposed. In the Home Life Building the staircases were saved owing to the fact that fireproof partitions prevented the fire from reaching them (see pages 19 and 20.)

STONE FRONTS. The damage done to the marble front of this building enforces the suggestions on pages 29 and 30. Projections for balconies, window-sills and caps, and other ornate finish are objectionable; they fall early in a fire and hinder the efforts of the firemen.

Stone fronts, especially if of marble, granite or other limestone, are liable to be severely injured by heat and water, and an entire facade may thus require to be replaced—a serious item in the case of ornamental or carved stonework.

Under the combined effect of fire and water all limestones show little more resistance than so much sugar.

To the extent, therefore, that the value of the stonework is a larger or smaller percentage of the value of the building, and according to whether the stone can be easily procured or must be purchased at monopoly prices, these facts should be taken into account, especially where there are exposures, in fixing the building line and insurance rate. It ought not to affect materially the line of insurance on the stock, however. Even where the building contains office furniture only, sufficient heat may pass out through the windows, as in the case of the Home Life Building and the Temple Court Building, to injure the stone lintels, ashlar, etc., (see the photograph of Home Life Building.)

VERTICAL PIPES for the use of fire department. A 4-inch or, better still, a 6-inch pipe, with outlets at each story, for the use of the Fire Department, as recommended on page 17, should be provided in all high buildings. In the case of the Home Life Building the tank was quickly emptied and no water could be secured on the upper floors, which were too far above the street to be reached by the fire department with hose. (The thread for coupling and the pipe should be the standard of the city fire department, $2\frac{1}{2}$-inch coupling, 8 threads to inch.)

WOODWORK AND OTHER TRIM. It is claimed that these can be fireproofed by processes which make them practically incombustible. Several buildings in New York have fireproofed trim, notably the Dun Building, Commercial Cable Building and the Queen Building. (See page 40.) It is claimed that some fireproof processes render the wood unfit for paint, varnish, etc., care should be exercised.

PARTITIONS. These should be fireproof, of well-burned porous clay. The hard-burned clay does not hold plaster so well as the porous kind. And they should in all cases pass through the flooring and rest upon the fireproof iron beams and arches, so as to thoroughly cut off the passage of fire from room to room; (see page 16.)

COMBUSTIBLE CONTENTS. This fire clearly demonstrated the fact that furniture, fixtures, merchandise, wooden flooring and trim will be as effectually cremated in a fireproof building as would the fuel in a stove. A number of the tenants in this building were uninsured, relying upon the fireproof character of the structure, overlooking the fact, strangely enough, that their property was not fireproof. (See page 21.)

It is my opinion that if the contents of these upper floors of the Home Life Building had been combustible merchandise solidly packed, instead of office furniture allowed as it was to burn out because the Fire Department could not get water to it, the iron framework of the building would have given way, and it is highly probable the upper part of the building would have fallen. One of the most practical Builders of New York, who has put up a number of fireproof buildings, once said that he believed a bad fire in the waist or middle of one of these structures would double it up like a jackknife. What would happen in case of the tumbling over of one of these enormous structures must be left to the imagination.

FIRE-EXTINGUISHING APPLIANCES. Tanks do not contain enough water to be of value for stubborn fires in these high structures unless an auxiliary source of supply is provided, and even in such case, as, for example, where a building is supplied with sprinklers and arrangements are made to attach a steam fire engine to a street outlet and so supply the system after the tank has been exhausted, it would be insufficient to accomplish much in the way of extinguishing a fire already under headway. A steam fire engine would supply only a given number of heads and would fall far short of supplying the sprinklers over a large territory. This is a possibility that seems to be very generally overlooked. It enforces the opinion that these buildings should not be constructed over 150 feet high and that mercantile and manufacturing buildings should not be constructed over 100 feet high, and that merchandise should never be stored higher than 100 feet above the grade.

PUMP-ROOM. It will be found that owners of fireproof buildings, and some underwriters, overlook the fact that the volume of water thrown to extinguish a fire is sufficient quickly to drown out the engine and pump room, put out the fires, expel the engineer and stop the elevators. If possible the pump and engine room should be so cut off from the rest of the building that it would not be flooded, no matter how much water should be thrown to extinguish the fire. This is not always easy nor possible, and where it is not, too much reliance should not be placed upon the independent pump system of the plant. Of course the outlet to the sewer and the sewer itself should be of sufficient capacity to carry off the water thrown by the fire department—probably as much as 4,000 gallons per minute, if say ten steam fire engines were working.

The staircases and elevators of a fireproof building should be enclosed in four brick walls, with fireproof doors protecting the communications with the main structure and outside windows as well in case neighboring buildings should project enough heat into the window openings to prevent the egress of inmates or the ingress of the firemen. Wire glass would be desirable for such windows.

CHANNELS FOR WIRES, PIPING, ETC. All of the electric-light wires, telephone wires, rising pipes, etc., should pass through the floors in this fireproof compartment, and the main standpipe, with hose outlets, should also pass up this shaft near the doors, so that it could be used to the last moment by the firemen. In most modern structures doors of rooms containing combustible material open directly opposite elevator shafts and stairways, and the smoke of a fire in any one room might render them quickly impassable.

STEAM JETS for Extinguishing Fire. These would prove, admirable for extinguishing purposes as they could be turned on by the engineer, with valves in the engine-room. A steam jet would quickly smother a fire in any one room or floor, and there ought always to be enough steam for the purpose on holidays and at night.

BOND STONES, STONE PIERS, ETC. Since the publication of the first edition of this work on the danger of bond stones in inside piers, fully explained on pages 29 and 30, a stubborn fire which occurred in the City of New York, January 18, 1899, in the Cammeyer Building, Sixth Avenue and Twentieth Street, has demonstrated the importance of employing only cast-iron for bonds and caps, and the great danger of using granite or other stone for the purpose. The accompanying photographs show the condition of brick piers five feet square after the fire was extinguished. The cracks extending from the cap stone through the brick work and bond stone below are plainly visible.

It will be apparent that if the fire had burned much longer these piers, whose integrity was vital to the structure, would have given way and let down the main girders. The cracks in the brickwork are due to the cracks first occurring in the cap and bond stones, the result of which was to throw the heavy superimposed weight upon a smaller section of the pier. If the cap and the bond had been of cast-iron the brickwork would have been uninjured.

This fire also demonstrated the truth of what has been claimed for cast-iron on pages 7, 8, 9 and 31. The cast-iron vertical supports, subjected to the same intense heat which ruined the brick and stone piers, although unprotected by fireproofing material, as they should have been, were entirely uninjured. Even 6-inch cast-iron columns supporting a girder in the rear were unbent and found to be thoroughly capable of carrying their superimposed loads after the fire.

It would have been better in this instance to have dispensed with the brick pier and its bond and cap stone and employed a circular cast-iron column, protected by eight inches of brickwork. This would have been less expensive than the large piers employed (five feet square) and would have been a material gain in the matter of saving space and light.

After a personal examination of this building, in which I had the benefit of the expert knowl-

HOW TO BUILD FIREPROOF. 85

edge of Chief Bonner, of the New York Fire Department, on whose judgment based upon many years of intelligent observation I greatly rely, I find nothing to take back in previous utterances on these two important subjects. Only cast-iron vertical supports should be used in cellars particularly, where they are subjected more than in any other place to dampness and the danger of rust. They should in all cases, however, be protected by coverings of fireproof material and they should be thoroughly tested for thin places caused by "floating cores" and other defects in casting. The thickness of the casting should be beyond all question as to factors of safety (never less than one inch for lower stories especially) and only round columns should be used.

The New York building law requires that no cast-iron column shall be of less average thickness than three-quarters of an inch ; nor shall it have an unsupported length of more than twenty times its least lateral dimensions or diameter. The law also requires that its bearings shall be faced smooth and at right angles to the axis of the column and when one column rests upon another column they shall be securely bolted together.

CAMMEYER BUILDING, NEW YORK.
FIRE, JANUARY 18, 1890.
DAMAGED BRICK PIER SHOWING EFFECT OF FIRE ON STONE CAP AND BOND STONES.
INTEGRITY OF PIER LOST.

CAMMEYER BUILDING, NEW YORK.
FIRE JANUARY 18, 1899.
DAMAGED BRICK PIER SHOWING EFFECT OF FIRE ON STONE CAP AND BOND STONES. INTEGRITY OF PIER LOST.

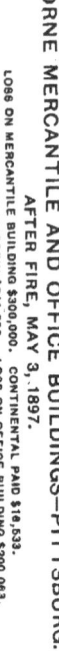

HORNE MERCANTILE AND OFFICE BUILDINGS—PITTSBURG.
AFTER FIRE, MAY 3, 1897.
LOSS ON MERCANTILE BUILDING $300,000. CONTINENTAL PAID $18,533.
LOSS ON STOCK CONTAINED THEREIN $741,250. LOSS ON OFFICE BUILDING $200,063.

HORNE OFFICE BUILDING PITTSBURG.

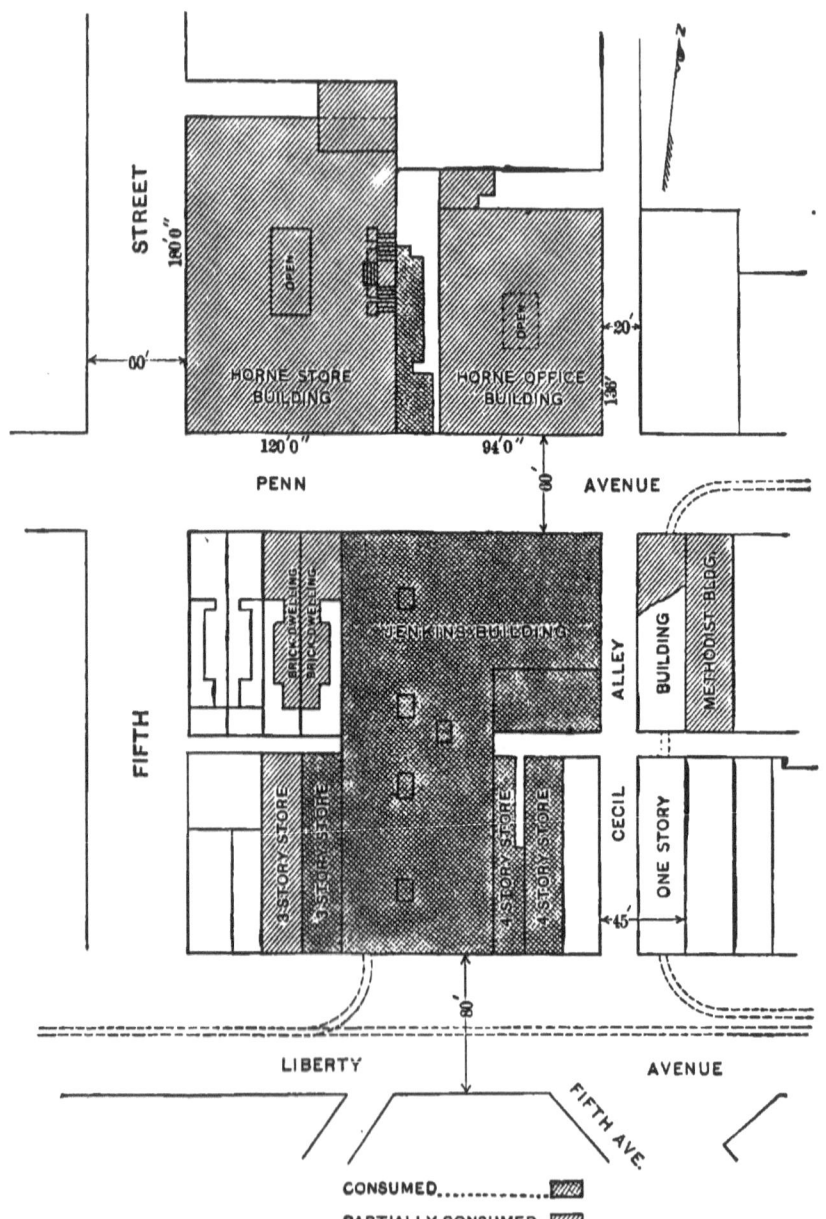

HORNE BUILDING (FIREPROOF) AND SURROUNDINGS, PITTSBURG, PA.
GROUND PLAN.
(Published by consent of THE ENGINEERING RECORD.)

FIRE JENKINS WHOLESALE GROCERY STORE, PITTSBURG.
FIRE MAY 3, 1897.
(Published by consent of THE ENGINEERING RECORD.)

FIRE IN JENKINS GROCERY BUILDING SHOWING HORNE BUILDINGS IN REAR.

HORNE BUILDING PITTSBURG, ENTRANCE.

HORNE BUILDING, PITTSBURG SHOWING INTERIOR AND WELL HOLE.

HORNE BUILDING, PITTSBURG.
SHOWING FIREPROOF FLOORS BROKEN THROUGH.

HORNE BUILDING, PITTSBURG—INTERIOR.

HORNE BUILDING, PITTSBURG.
SHOWING COLUMNS STRIPPED WELL HOLE, ETC.

HORNE BUILDING, PITTSBURG, UPPER FLOOR AND SKYLIGHT.

HORNE BUILDING, PITTSBURG—INTERIOR.

HORNE BUILDING, PITTSBURG.
SHOWING FALLEN TANK, ETC.

HORNE BUILDING, PITTSBURG ROOF ETC.

HOME LIFE "FIREPROOF" BUILDING, NEW YORK.
FIRE DECEMBER 4, 1898.
SHOWING BUILDING BEFORE THE FIRE.

HOME LIFE "FIREPROOF" BUILDING, NEW YORK.
FIRE DECEMBER 4, 1898.
SHOWING BUILDING AFTER THE FIRE.
FIRE DAMAGE APPRAISED $200,000. OF WHICH THE CONTINENTAL'S SHARE WAS $40,000.

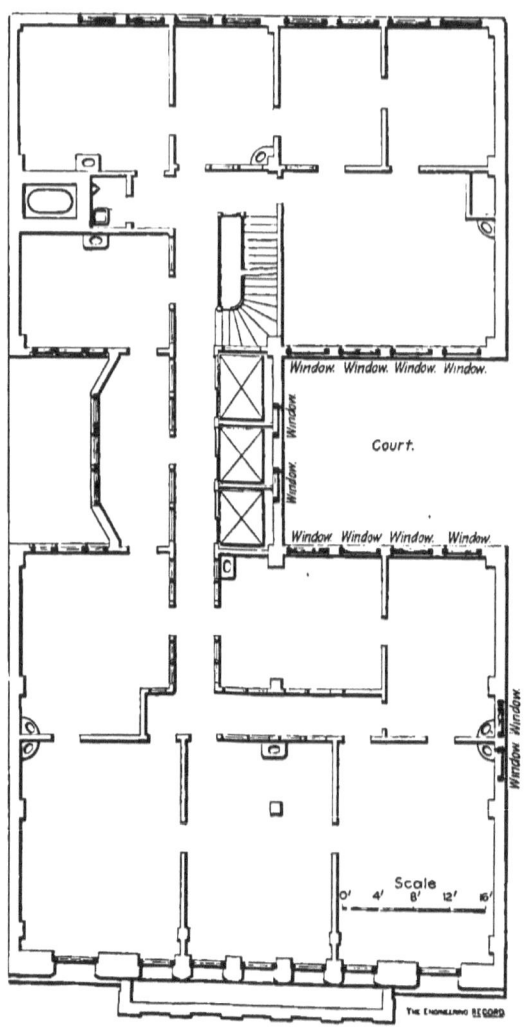

GROUND PLAN OF HOME LIFE BUILDING,
SHOWING COURT, ETC.
(Published by consent of THE ENGINEERING RECORD.)

HOME LIFE BUILDING, SHOWING COURT,
WHICH BY INCREASING DRAFT CONTRIBUTED TO THE DAMAGE.
(Published by consent of THE ENGINEERING RECORD.)

HOME LIFE BUILDING. STONE FRONT AFTER FIRE.
(Published by consent of THE ENGINEERING RECORD.)

ROOM IN HOME LIFE BUILDING AFTER FIRE.
(Published by consent of THE ENGINEERING RECORD.)

MANHATTAN SAVINGS BANK BUILDING.
FIRE NOV. 4, 1885.   PAGE 20.
LOSS $225,000.   CONTINENTAL INS. CO., PAID $20,355.

INSTRUCTIONS FOR CONSTRUCTION

— OF —

FIRE DOORS AND SHUTTERS

— ADOPTED BY —

New York Board of Fire Underwriters

MARCH 17, 1897.

Placing your fire insurance in some companies is like

KEEPING YOUR VALUABLES

IN A PASTEBOARD BOX

INSTEAD OF A MODERN SAFE

when you have the choice of either at the same price.

Get a CONTINENTAL policy and you are sure of absolute indemnity at fair rates.

An old established AMERICAN COMPANY whose fixed policy, financial strength, progressive management and fair treatment of policyholders, are guaranteed by its past record.

"Insure In an American Company."

CONTINENTAL FIRE INS. CO.,
46 Cedar Street, New York.
Rialto Building, Chicago, Ills.

INSTRUCTIONS FOR CONSTRUCTION

OF

FIRE DOORS AND SHUTTERS.

Communications between buildings should be cut off by fire doors each side of opening, as follows:

There should be no woodwork or furring about the opening; the doors should be made of two thicknesses (except where openings are made larger than called for in the Standard. See page 11) of 1-inch narrow, tongued and grooved, soft white pine boards (free from sap, pitch or moisture of any kind), laid diagonally (both sides) and nailed with wrought iron nails driven through and clinched, then covered on both sides and edges with 10x14-inch sheets of "bright I. C." tin, except where doors are exposed to an atmosphere liable to cause rust (then TERNE PLATES should be used in place of tin), joints flat-locked and securely nailed under laps (barbed wire nails to be used, 1½ inches in size, to be driven

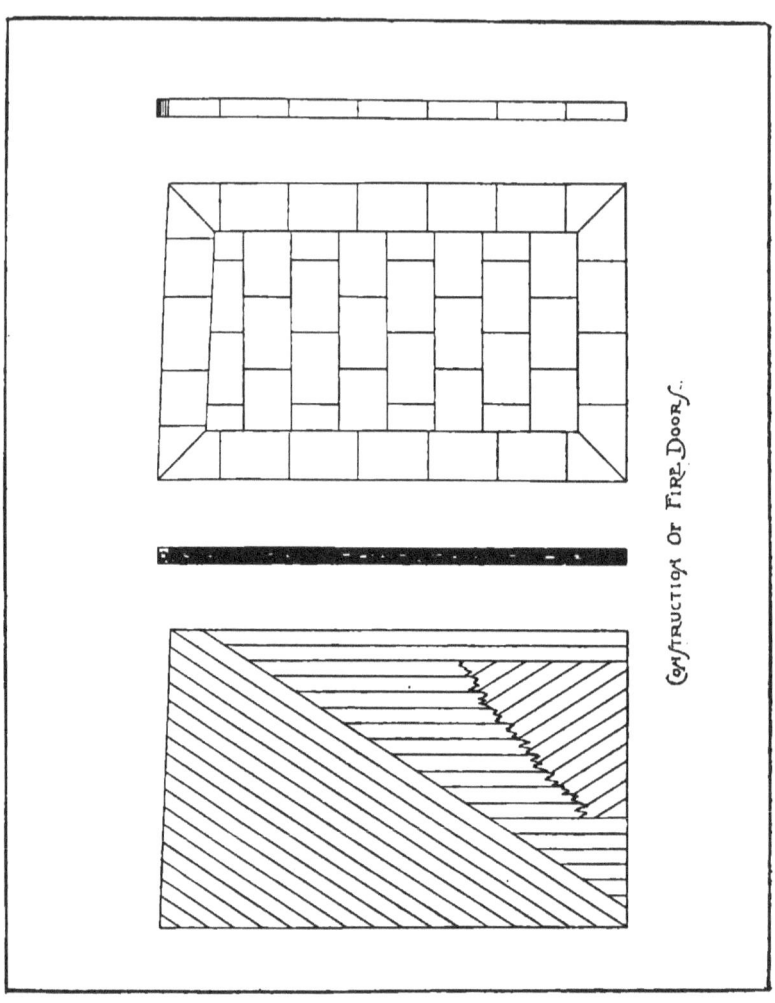

inches apart; for shutters 1-inch barbed wire nails should be used), but not soldered.

The plugging of walls with wood or lead, or the use of lag screws will not be permitted; tracks or other fittings put up in this manner, will not be approved.

AUTOMATIC DOORS.

It is preferable that all communicating doors be arranged to close automatically in case of fire, and where a building fully equipped with sprinklers communicates with one not so equipped, the doors at the communications must be arranged to close automatically. [See drawing on page 4.]

SLIDING DOORS.

[See drawing on page 4.]

When sliding doors are used, they must be of sufficient size to lap 3 inches over the opening at sides and top, and they must be hung to suitable steel trolley tracks, and the brackets, hangers and all other fittings in connection with tracks and doors should be made of wrought or malleable iron, and hangers should be firmly bolted to and through doors by carriage bolts, and

Where walls are 16 inches or less in thickness, then fastenings of tracks, hinges and other fittings for carrying doors or holding

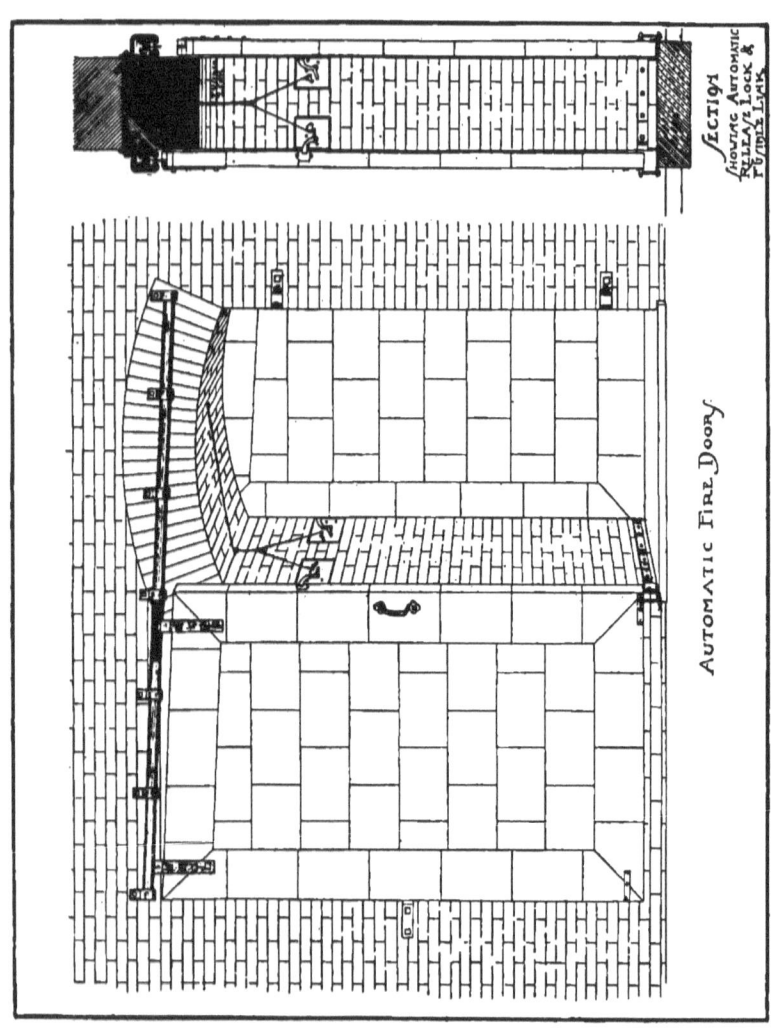

the same fast to brickwork, should be bolted to and through same with necessary washers and nuts, and

Where walls are over 16 inches in thickness, brackets for holding tracks should be firmly bolted to the brick wall (both sides) by expansion bolts ½ x 6 inches, with not more than 20 inches space between each bolt. The tracks should be placed at an incline not less than ⅜ to ½ inch to the foot, and at such height that the door, when closed, will rest firmly on the sill, and at the lower and upper corner about 6 inches from top and bottom, iron binders (counter-sunk) must be bolted to brick wall and jambs by expansion bolts ½ x 6 inches, to keep doors in place, and when closed, to hold them close to the wall, and a guide also bolted to brickwork should be placed at the other lower corner, as shown in drawing on page 4. Handles should be placed on doors (counter-sunk) on the inside and pull handles on the outside. [See drawings on pages 17 to 19 for fittings.]

When required, a light framework of slats must be built outside of doors to prevent piling of stock, etc., against same.

In order to allow the doors to run free and not chafe against the wall, the roller part of guide should be set about ½ inch from door when hung and open; then a wedge-shaped piece of iron (as shown in drawings on pages 4 and 17) should be screwed to door near the lower corner in such a position as to bring the wedge between door and roller guide, so as to clamp the door tight against the wall when closed.

If chafing strips are used on sliding doors, they should be

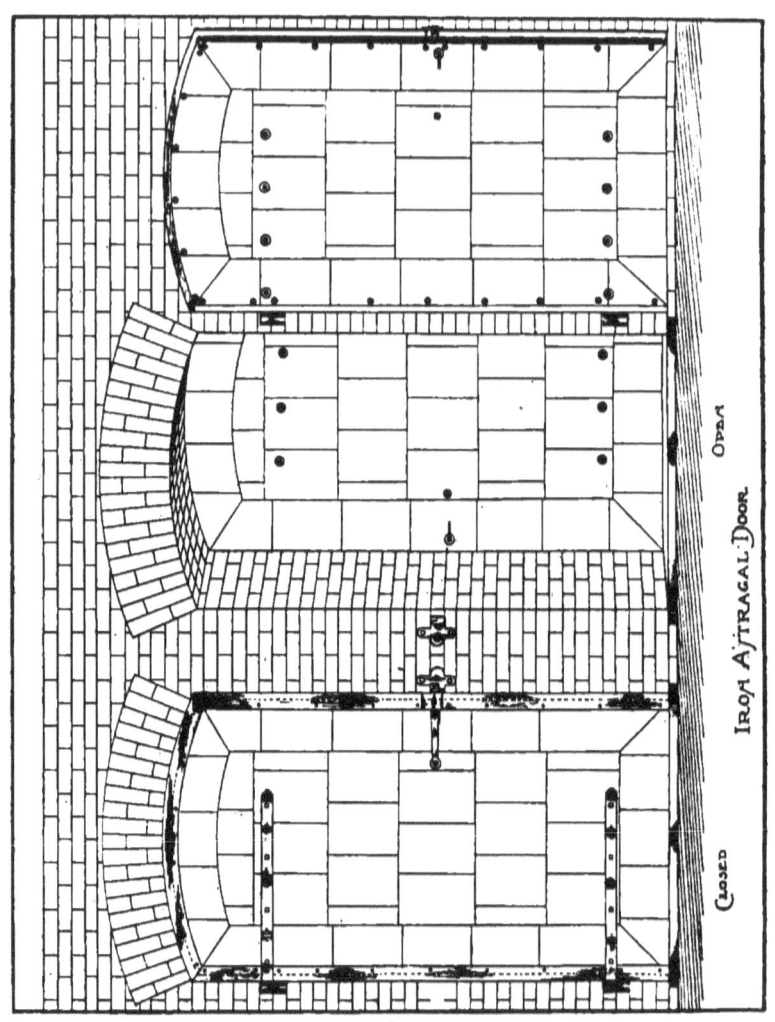

screwed to the inside part and run across door about 5 inches shorter at each end than the width of doors.

SWINGING DOORS.

When swinging doors are used they must be hung on iron holdfasts built in wall, or hooks bolted through the wall, or by wrought or malleable iron plates firmly set by expansion bolts, or hinged to iron rabbeted jambs [see drawing on page 10] securely bolted to brickwork, and the hinges should be of wrought iron $\frac{1}{4}$ to $\frac{3}{8}$ inch thick by $1\frac{1}{2}$ to 2 inches wide, extending two-thirds across each door, bolted to and through same by carriage bolts, the heads of bolts resting against washers outside of tin, and nut to screw against hinge; doors when closed should be fastened by heavy wrought iron drop-latch bolted to and through doors in the same manner as with hinges; catch to receive latch must be securely fastened to brickwork. If bolts are used then iron plates with opening for bolts to slide in must be bolted to brickwork.

All swinging doors to close in jambs flush with face of wall. [See drawings on page 6 and 8.] Where doors close in jambs without rabbet, they must have iron astragal $2\frac{3}{4}$ inches wide, $1\frac{1}{2}$ inches on doors, all around outer edge of doors, fastened by bolts set 8 inches apart, bolted to and through same, and lapping over brickwork $1\frac{1}{4}$ inches. [Drawing on page 6, shows astragal door. Drawing on page 8, shows rabbet built in brickwork. Drawing on page 10, shows rabbet made by angle iron.]

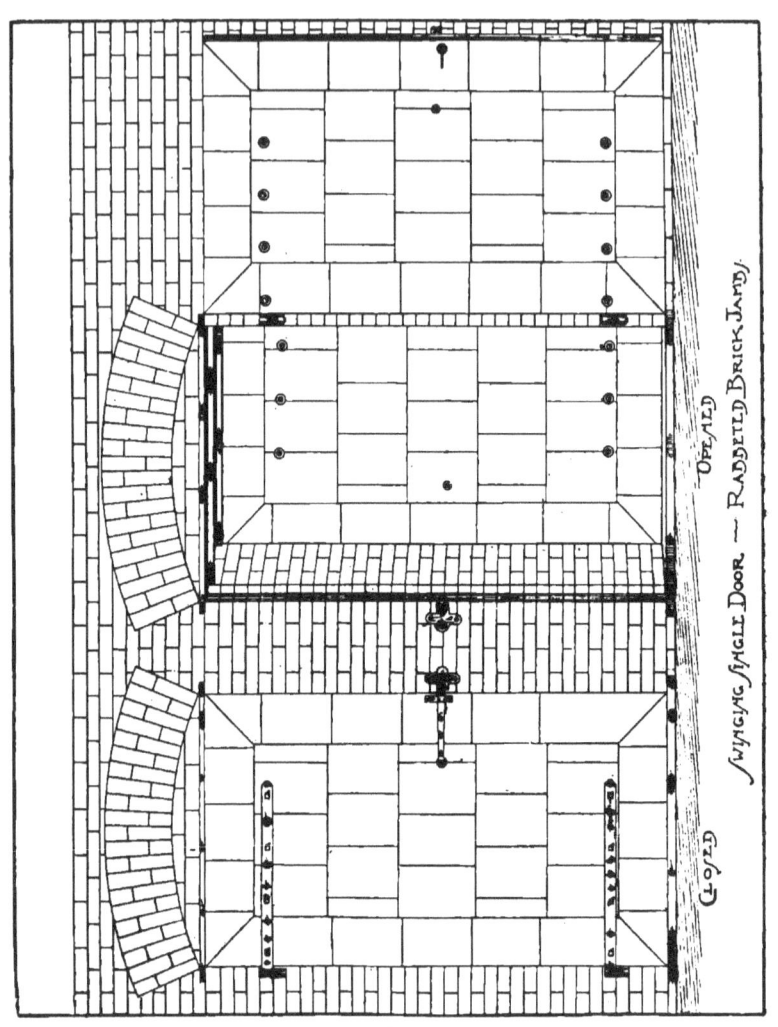

SWINGING SINGLE DOOR. — RABBETED BRICK JAMB.

RABBETED DOORS.

When the doors are made in two parts the edges of the doors where they come together should be rabbeted, the single boarding of each door extending over and fitting in that of the other about 1 inch [see drawing on page 12], so as to make a close joint when closed; these doors to be hung and hinges put on the same as swinging doors, and the door that is closed first should be fastened by bolts on the inside, top and bottom, let into sills and lintels; and on the door that closes last, there must be placed a wrought or malleable iron drop-latch (as shown in drawings on pages 12 and 18), with latch-guide firmly bolted to and through door by carriage bolts, and with catch placed on the opposite door in such manner that the door can be opened from either side.

SILLS.

All sills must be of stone or iron and rest on the solid brick wall with highest side at least 1½ inches above the floor and to be full width of the wall, and set in the brick jambs 2 inches and extend at least 3½ inches beyond the face of wall, both sides, so that inclines even with top of sills may be used and not interfere with the closing of doors. [See drawing on page 16.]

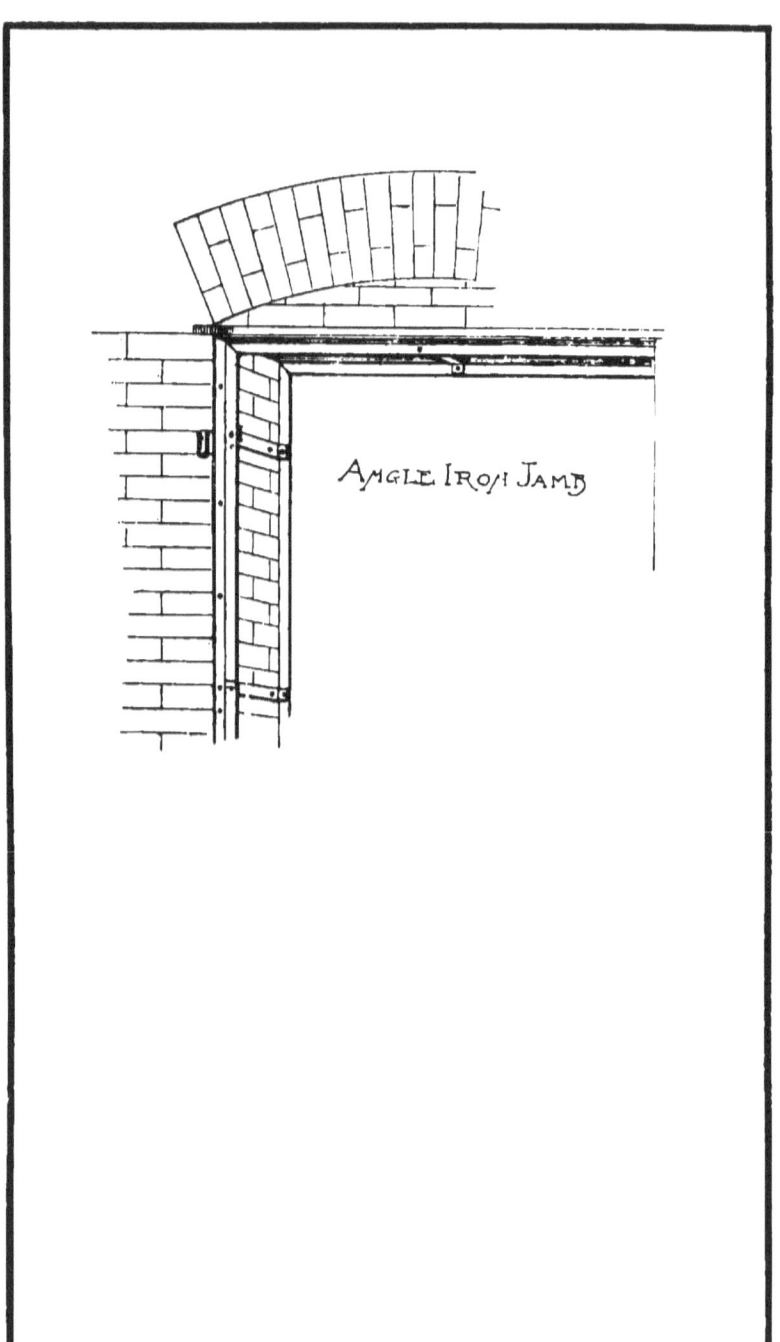

LINTELS.

Lintels should be of stone or iron, but if wood lintels are found in old arches the same must be protected by tin, put on the same as on doors, edges flashed in brickwork.

OPENINGS.

When openings are over 4 feet in width and 7 feet in height, then doors must be increased in thickness in proportion to the increased size of said openings, but should not be larger than 6x8 feet, and should be made of three (instead of two) thicknesses (as shown in drawing on page 2, and Section on page 17).

FIRE SHUTTERS.

All batten fire shutters should be made the same as rabbeted doors, except two thicknesses of ⅞-inch or not less than ¾-inch narrow, tongued and grooved, soft white pine boards should be used, the same as for fire doors, and the edges covered with galvanized iron, in strips about 9 inches in width, forming a border to cover edges and lap over on each side of shutter about 3½ inches, the edges of same to be turned up to receive Terne Plate (in place of tin) in order to form a proper lap joint; the Terne Plate to have

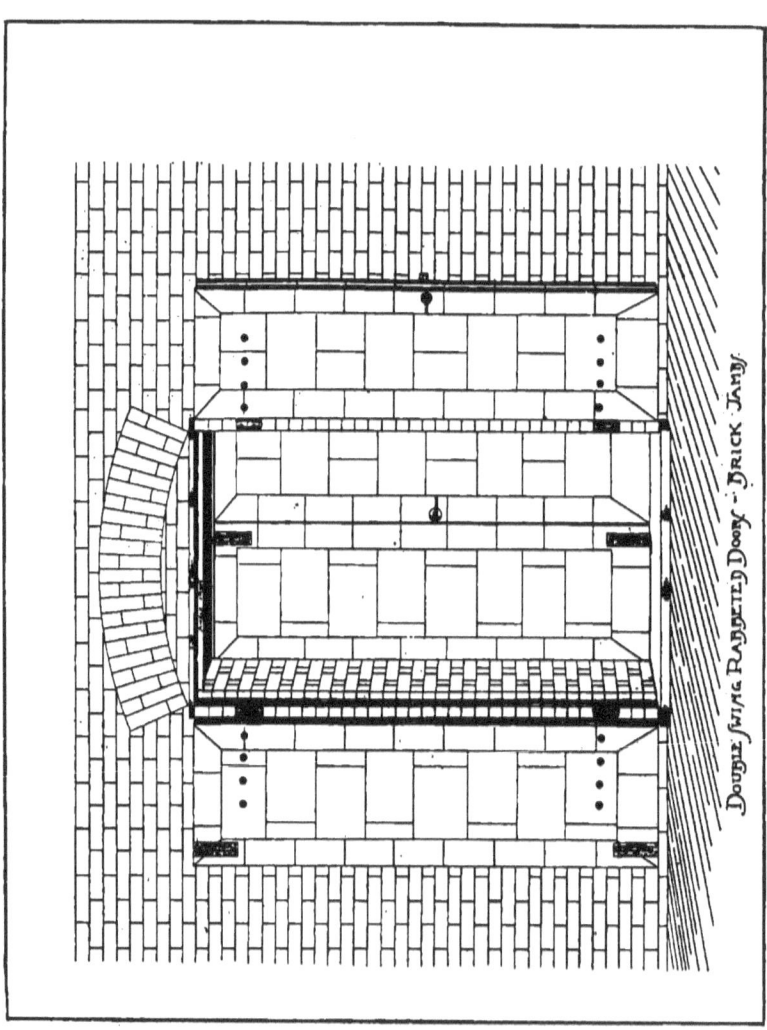

Double Swing Rabbeted Doors - Brick Jamb.

no vertical joints except with border, and the galvanized iron border to have joints, one at each corner; and all cross-lap joints should be made in such a manner that water will not work in back of plates to wood (the galvanized iron to conform as near as possible to the grade of I. C. tin).

Where shutters come together, they should be rabbeted, and provided with strap hinges to run two-thirds across each, bolted to and through same by carriage bolts, and hung to wrought iron eyes built in brickwork, or by bolts passing through walls, held by nuts and washers, so that they can close inside of brick jambs and be fastened to iron cross-bar of not less than $3/8$ to $1/2$ inch thick by $1\tfrac{1}{2}$ inches in width, ends let in brickwork. [See drawing on page 14.]

In cases where shutters cannot be properly placed to close inside of brick jambs, then they must be made to lap 3 inches over the window openings (at sides and top), and arranged to close tight on stone or iron sills.

The fastenings or latches at windows [see drawing on page 19], above the first floor, should be so arranged, when required, that they can be opened from the outside [see drawing on page 14].

Where standard wood, metal covered, shutters are not required, and

IRON SHUTTERS

are permitted, the said iron shutters should be made as follows:

If the window openings are of ordinary size, say 4 feet 6 inches

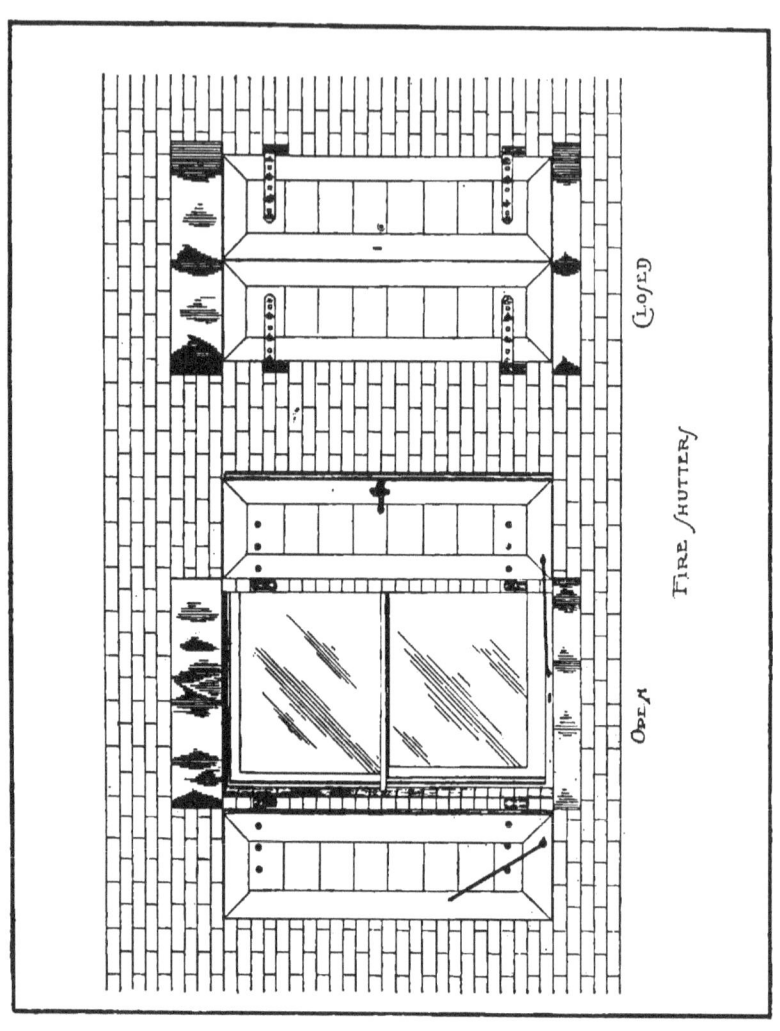

by 8 feet 6 inches to 9 feet, the frame should be of 3-16 angle iron 1½x1½ inches.

For openings of from 6x10 feet, the frames should be not less than ½ inch angle iron 2x2 inches. For larger openings than the above, frames should be increased in proportion to the size of shutter, and be covered with not less than No. 16 blue annealed iron.

For all shutters over 8 feet high, at least three eyes shall be provided on each side, and if hung to iron hanger frames, at least three anchors to brickwork, on each side, shall be provided.

All iron shutters should be hung to suitable iron hanger frames, or to iron eyes securely fastened in the walls, and so arranged that when closed, the shutters will be at least 4 inches free and clear of all woodwork.

Shutters should be kept well painted.

ADDITIONAL INSTRUCTION.

In all the cases above-mentioned, work will be subject to the approval of the Inspector of this Board, and should there be any doubt as to the method of doing the work or any change or modification, further instruction will be given at this office.

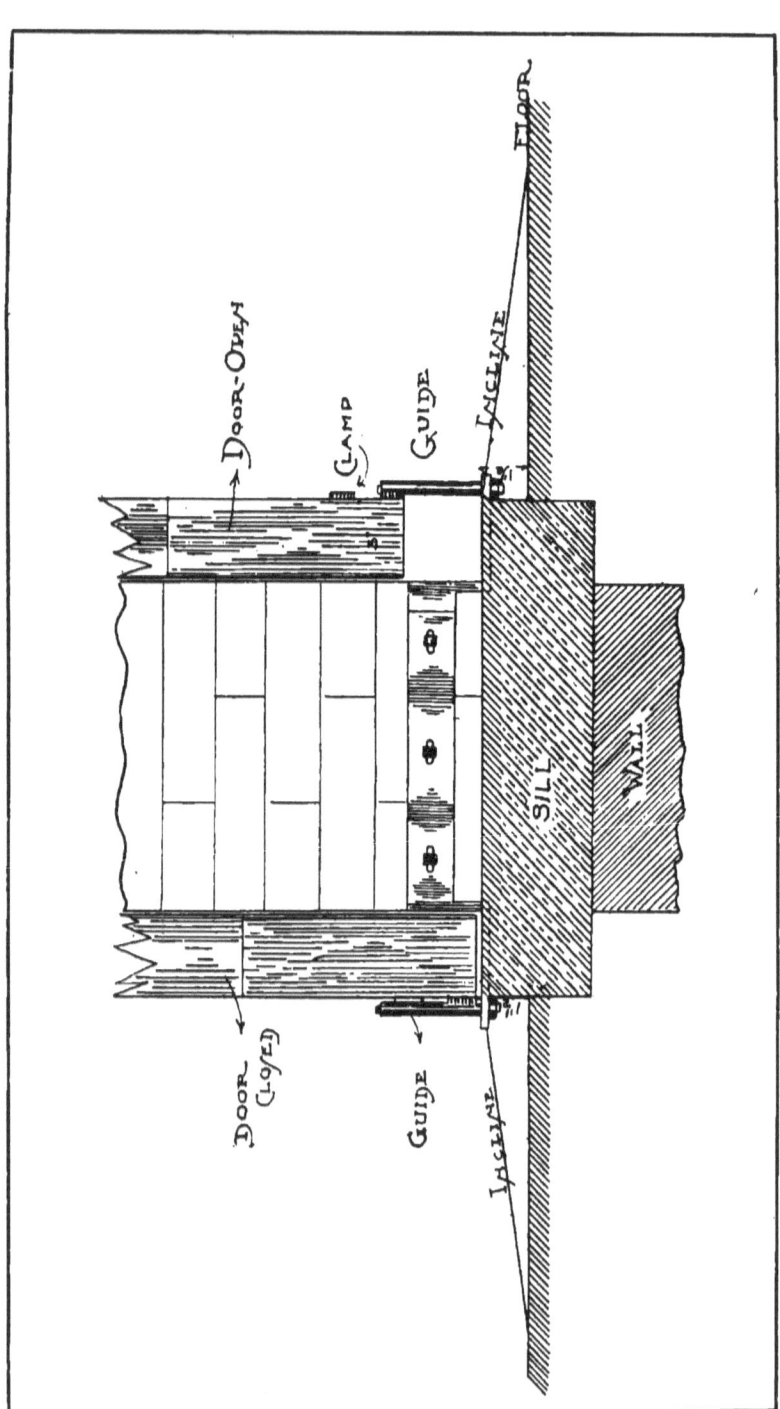

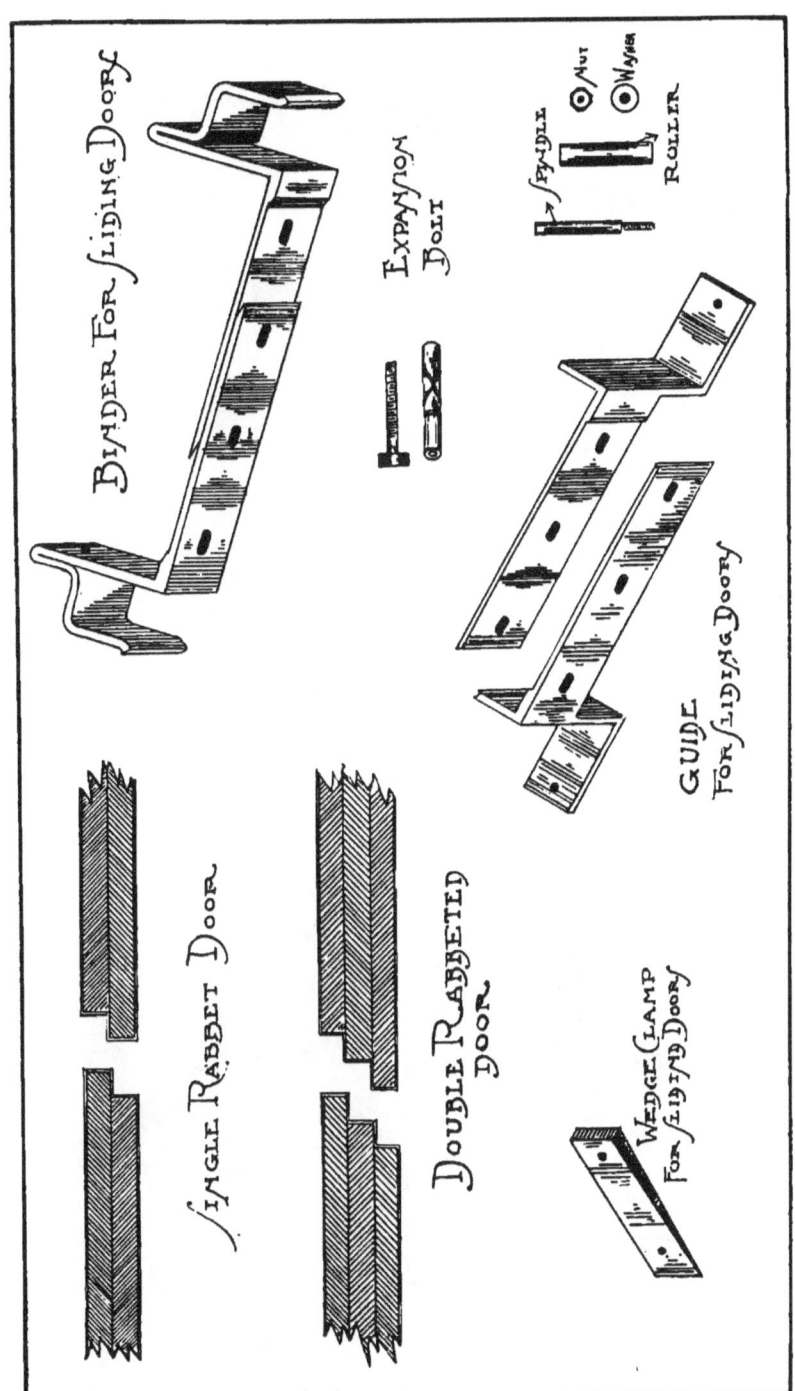

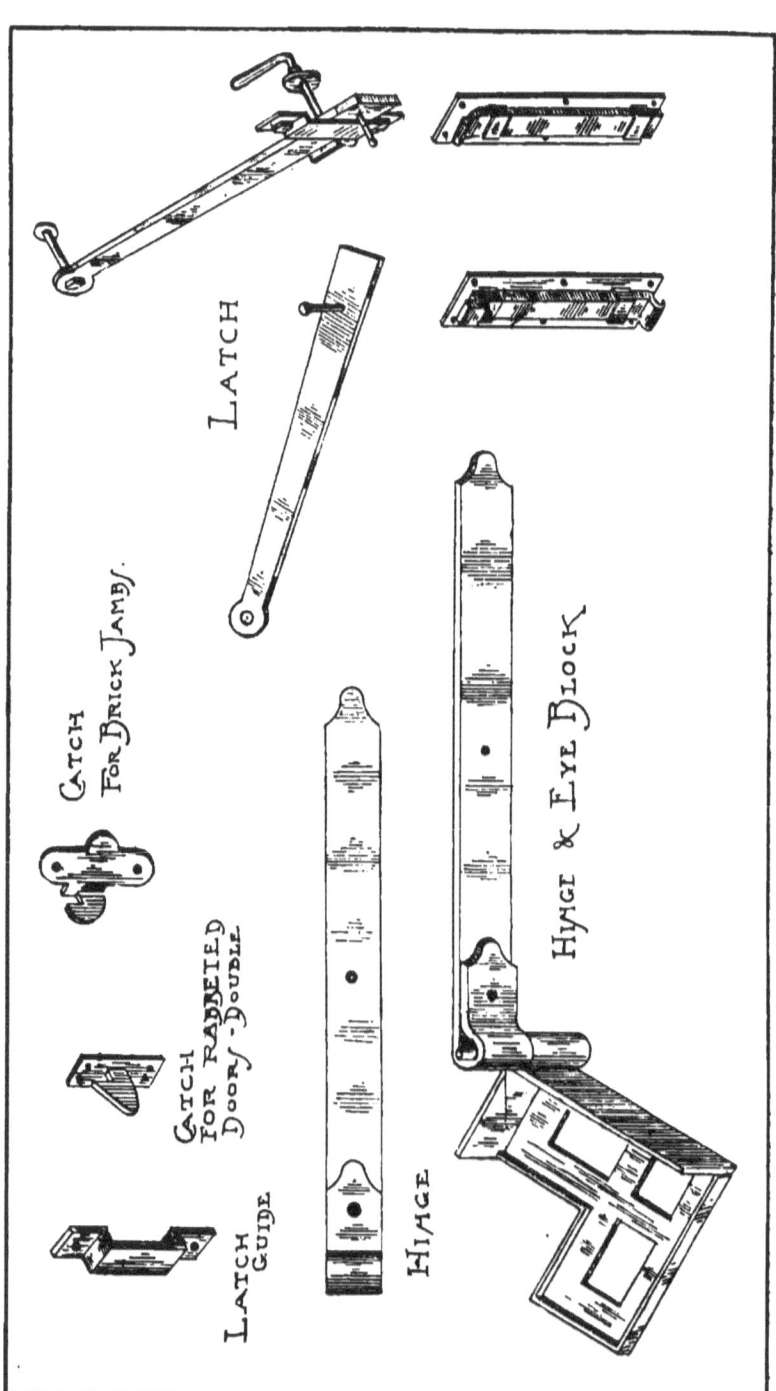

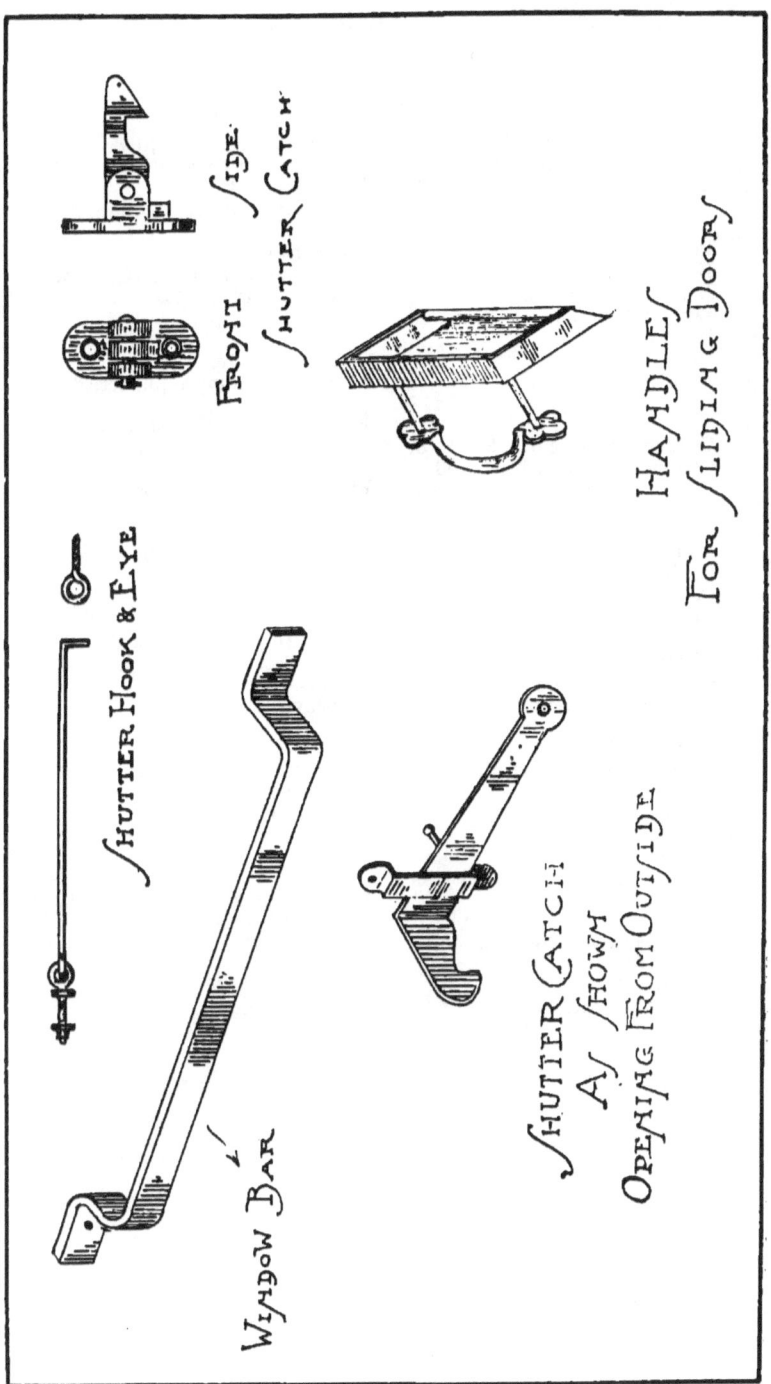

INDEX.

A.

American Exchange Bank, 30.
Arches between beams, 36.
Automatic Sprinkler Pipes, 73.

B.

Beams, 35.
Boiler room, 44.
Bond stones, 29, 83.
Bonner, Chief, 85.
Braces, iron, 76.
Brick, 25.
Burnley Mills, England, 34.

C.

Cammeyer b'ld'g fire, 83, 86, 87.
Capitol at Albany, N. Y., 33.
Cast iron, 7, 9, not liable to rust, 8, 31.
Cellars, dampness in causes rust, 11.
Ceiling, wooden, 72.
Cement, 9.
Channels for pipes, wires, etc., 82.
Chimneys, 45, 46, 47.
Closets, 70.
Columns arranged for stripping and examination, 9.
Communications, 48.
Conductivity, 12.
Contents of fireproof buildings, destructibility, 21.
Conflagrations, 22, 23.
Courts as flues, 54.
Curtain walls, 25.
Cut offs, gas and electricity, 62.

D.

Doors between buildings, 48.
Doors, fireproof, 53, 75.
Dry rot, 71.
Dynamo rooms, 45, 62.

E.

Electric lighting, 45, 70, cut offs at street for firemen, 62.
Elevators, 7, 42, 67.
Enclosing walls, 22, 68, thin, 25, 27.
End construction, 36.
Expansion of iron, 11, 12, 34, of masonry, 12, 13, 14.

F.

Fire escapes, 52.
Fire extinguishing appliances, 62, 73, 81.
Fireproof doors, 48, 53, 75.
Fireproof shutters, 53, 75.
Fireproof wood, 40, 80.
Fireproofing iron members, 7, 32.
Fireplaces, 45.
Floor arches, 36.
Floor boards, 40, 67, 76.
Foundation, 67.

G.

Gas, cutting off supply main, 62.
Glass windows, 6, wire glass, 54.
Girders, 35.
Guastavino arches, 77.

H.

Hallway, fireproof, 48, 49, 50, 51.
Heat of burning building, 22, 27.
Home Life Building, 75.
Horne Building, 6, 21, 41.

I.

Insurance companies interest in fireproof construction, 57.
Iron columns, 85.

L.

Luxfer prisms, 41.

INDEX.

M.

Melting points, 22.
Merchandise in high buildings, 16
Mills, Burnley, 34.
Mills Building, 7.
Mill construction, 63.

N.

New York Board rules for fireproof doors and shutters, 109.
Non-fireproof buildings, 63.

O.

Oil fires, sand for, 73.
Openings through floors, 7.
Outside stairways, 52.

P.

Parapet walls, 69.
Paint, 10.
Partitions should be fireproof, 16, 80.
Patent floor arches, 37.
Piers, 29, 84.
Pillars, stone, 30.
Post, George B., Col., 32.
Protected iron work, 7, 32, 35.
Pump room, 82.

R.

Recapitulation, 25, 59.
Rivets, 11.
Roof, 16, 43.
Rubbish, 71.
Rust, danger of, 8, 10.

S.

Safes, iron fireproof, 77.
Sand to extinguish fires, 73.
Sawdust spittoons, 71.
Schedule insurance rates, 63, 66.
Sheathing, wooden, 72.
Skeleton construction, 28.
Skylights, 42, 74.
Shims, 31.
Shutters, fireproof, 53, 109, self-closing, 55.

Side construction, 36.
Slate roofs, 43.
Slow burning construction, 63.
Sprinkler pipes, 73.
Stairways, 7, 41, 67, 78, 82, stone treads, 19, 20.
Stand pipes, 17, 79.
Standard building, 66.
Steam jets, 83.
Stone bonds, 29.
Stone fronts, 30, 78.
Stone treads, 19, 20, pillars, 30.
Stores and Warehouses, 63.
Summary, 25.
Steel, 8.
Steel rivets, 11.

T.

Tanks, 17, 52, 69.
Terra cotta arches, 38.
Tests of material, 56.
Templates, 52.
Tie rods, 36.

U.

Underwriters, interest in fireproof construction, 57.
Underwriters' doors and shutters, 109.
Universal Mercantile Schedule, 66.

V.

Ventilating shafts, 40.
Vertical stand (water) pipes, 17.

W.

Walls, thickness of etc., 22, Thin, 25, 27.
Watchman for high buildings, 18.
Waste, 71.
Water tanks, 17, 52, 69.
Water proof floors, 39, 70.
Well holes, 41, 73.
Western Union Building, 12.
Window frames, 78.
Wire glass, 54.
Wooden sheathing on side walls, ceilings, etc., 72.

WHY TO INSURE IN AN AMERICAN COMPANY.

AMERICAN COMPANIES ARE THE LARGEST;
of the six companies (including the "Continental") reporting OVER TEN MILLION DOLLARS IN ASSETS, only one is foreign, and its U. S. assets are less than those of the "Continental".

AMERICAN COMPANIES ARE THE STRONGEST;
of the five companies (including the "Continental") whose reports show a surplus to policyholders EXCEEDING FIVE MILLION DOLLARS, none are foreign.

COSTS NO MORE.
Why patronize foreigners when you can get the same thing at the same price from fellow-countrymen?

GIVES BUSINESS TO THOSE WHO GIVE YOU BUSINESS:
Stockholders of the American Companies are their partners and as they are distributed throughout the United States, they are doing business with you.

PROFIT, IF ANY, REMAINS IN THIS COUNTRY,
contributing to the general prosperity, which in turn benefits YOU.

THE
CONTINENTAL
INSURANCE
COMPANY
of New York.

WHY TO INSURE IN THE CONTINENTAL.

Is an American Company, owned by Americans and managed by Americans.

Does business under the Safety Fund Law, making its policy "Conflagration Proof."

Assets ($11,599,011.) and surplus ($5,901,328.) to policyholders are larger than those in the U. S. of any foreign company.

Paid in full all losses incurred in the great Chicago and Boston conflagrations.

Since organization its loss payments to policyholders exceed Forty-Seven Millions of Dollars.

You secure, if desired, the advantage of inspection by experienced men, and will be furnished on request with information regarding safe construction of buildings, etc.

Prompt attention to loss adjustments ensured by the organized force of travelling men which the Company's large business enables it to maintain to cover every section of the country and which a smaller company could not afford.

Organized in 1852, its fifty years of successful business prove its financial strength, conservative management and fair treatment of policyholders.

www.ingramcontent.com/pod-product-compliance
Lightning Source LLC
Chambersburg PA
CBHW022135160426
43197CB00009B/1292